THE BOATMAN

The Boatman

HENRY DAVID THOREAU'S
RIVER YEARS

ROBERT M. THORSON

Harvard University Press

Cambridge, Massachusetts & London, England

First Harvard University Press paperback edition, 2019
First printing

Library of Congress Cataloging-in-Publication Data

Names: Thorson, Robert M., 1951– author.
Title: The boatman : Henry David Thoreau's river years / Robert M. Thorson.
Description: Cambridge, Massachusetts : Harvard University Press, 2017. |
    Includes bibliographical references and index.
Identifiers: LCCN 2016046669 | ISBN 9780674545090 (cloth : alk. paper) |
ISBN 9780674237414 (pbk.)
Subjects: LCSH: Thoreau, Henry David, 1817–1862. | Stream conservation—
    Massachusetts—Concord River Valley. | Concord River Valley (Mass.)—
    Environmental conditions.
Classification: LCC PS3057.N3 T486 2017 | DDC 818/.309—dc23
    LC record available at https://lccn.loc.gov/2016046669

Musketaquid. Replica of Thoreau's most well-known boat on the shore of Fairhaven Bay.
September 2016. Boat loan courtesy of Concord Museum. Photo by Juliet Wheeler.

To all the dedicated conservation groups in Thoreau country,
especially OARS, the watershed organization for the
Assabet, Sudbury, and Concord Rivers

THE EARTH IS OUR SHIP, AND THIS IS THE SOUND OF
THE WIND IN HER RIGGING AS WE SAIL.

Journal, January 2, 1859

Contents

Figures

Preface

FOURTEEN YEARS AGO I walked down to the basement archive of the Concord Free Public Library to meet its curator, Leslie Perrin Wilson. Before being buzzed through the heavy glass door, I saw Henry Thoreau's brass surveying compass mounted on its original wooden tripod within a tall display case. Using this time-tarnished instrument, Thoreau helped dozens of nineteenth-century factory owners, land speculators, local governments, and farmers develop their properties during the nineteenth-century makeover of the Concord River Valley.[1]

Sensing my interest in surveying, Leslie disappeared into the vault for a few minutes. She returned with what looked like a tight roll of fabric wallpaper that was frayed along the edges, stained with a bad glue job, and smudged with what I later learned was meadow mud. This was Thoreau's handmade, cloth-backed, seven-foot-long field map of the Concord River. To lay it flat on the glass-topped library table, we used black-velvet beanbags to hold back its curled ends. We read it like an ancient scroll.[2]

"What do you make of this?" she asked, or something to that effect. Beyond the obvious fact that "this" was a tightly rolled map of some sort, I had no better answer at the time.

Since then I've returned to the archive many times to be buzzed through the security door, don white cotton gloves, borrow a soft number-two pencil, unroll the scroll, and read it with a magnifying glass. It's the Rosetta Stone for an overlooked piece of Thoreau's biography, the nautical chart of his boatman's life, and an important clue to a turning point in American environmental history. Surprisingly, the scroll map is anonymous.[3]

Graphically, the map's a mess. An inked double line widens and narrows to show the river channel flowing downstream with graceful meanders, sharp zigzags, ragged crenulations, and fjord-like reaches. Between the lines are depth soundings taken from his boat with a long birch pole, ranging from two and a half feet below weedy gravel bars to nineteen and a half feet in deep

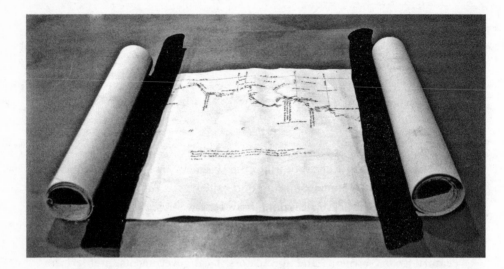

FIGURE 1.   Scroll map. Thoreau's compilation map for his river project is glued to a cloth scroll more than seven feet long. Document 107a of online exhibit of Thoreau's surveys. Courtesy of the Concord Free Public Library. Original size is 91 by 15 inches. A high-resolution scan of this portion shown in Figure 22.

bedrock pools. Each measurement is referenced to a notch in a willow tree on land owned by his friend William Ellery Channing, which is referenced to the cornerstone of a stairway built by his neighbor Edward Hoar, which is referenced to an iron truss for a railroad bridge built by the Boston and Fitch-burg Railroad, which is referenced to the geodetic sea level of the Atlantic Ocean. Though proper geographic names are beautifully rendered in calli-graphic style, the names of familiar personal places such as "boat place" and "swim hole" are scribbled sideways or upside down. Some words are nearly microscopic. There are three colors of ink and several shades of pencil, indi-cating a multistage production history. The whole thing is a cross between a technical illustration and a journey map of Thoreau's private life: similar to but far richer than its literary counterpart, the famous bathymetric map of Walden Pond.

The base map for Henry's scroll was created twenty-five years ear-lier, in 1834. Thoreau's version dates to July 1859, less than three years be-fore his untimely death in May 1862. Twelve years later in 1874, his younger sister, Sophia, donated the scroll to the Concord Library as part of a bundle of maps and personal papers her brother kept in their nearby family house at 255 Main Street. Likely this was in his third-floor, two-room attic suite, guarded by the three women of the house. This self-described "sanctum" or "chamber" was Henry's bedroom, office, library, studio, writing retreat, herbarium, and natural history museum during the final and most productive decade of his life, the only one for which he kept a regularly dated journal. The garret's western window overlooked the river, which was close enough for him to be awakened at night by the torchlights of men spearfishing and by bullfrogs croaking at water's edge. Morning fogs poured through the western window like cotton. Town bustle poured in from the east. Across the street and beyond William Ellery Channing's garden was his boat "harbor" in the willows.[4]

The Thoreau who drafted the scroll map in 1859 is marginalized in biography. He was not the literary stylist who gave us *Walden,* arguably the most important work of American literary nonfiction, but the sole author of an unfinished and unpublished scientific report on a greatly transformed but still achingly beautiful river system. Not the political dissident known for "Civil Disobedience," an essay inspiring generations of nonviolent protestors, but a private technical consultant working on a statewide environmental controversy that spanned four years and required four separate acts of the Massachusetts state legislature. Not the patron saint of wilderness known for his essay "Walking," but someone who favored rowing and sailing a sluggish river that was "dammed at both ends and cursed in the middle."[5]

The mapmaker was an older, wiser scientific genius whose fatal illness would manifest itself only one year later. He was the valley's most widely respected naturalist, admired for his intimate knowledge of the ways and means of local waterways. Someone listed as a civil engineer on the official town map. Someone qualified to oversee Harvard's science curriculum. Someone so well known in a bustling town of more than two thousand inhabitants that he had recently received an unaddressed postcard with his surname misspelled. The legal head of a household who enjoyed the cracker-barrel politics of American life during his daily trips to the post office. The forty-two-year-old chief executive officer of his family's successful black-lead manufacturing business that operated in the shadow of their more visible pencil business. A self-described "boatman" whose favorite flowing stream was the most disrupted river reach of the region, the lowermost three miles of the Assabet.[6]

FIGURE 2. **Dunshee ambrotype.** Photograph (ambrotype) of Henry David Thoreau, by Edward Sydney Dunshee, New Bedford, Massachusetts, 1862. Gift of Mr. Walton Ricketson and Miss Anna Ricketson (1929), THO033B. Courtesy Concord Museum.

On June 4, 1859, this little-known Thoreau was hired by the River Meadow Association, a seven-town coalition of farmers demanding removal of the downstream factory dam in Billerica. They claimed it was back-flooding up to fifteen thousand acres of their rich alluvial valley and ruining their agricultural economy, which was based on meadow hay, and to a lesser extent, cranberries. This "flowage controversy" was arguably America's first major environmental debate over dam removal, a veritable class-action suit with more than five hundred petitioners that culminated a half-century of legal conflict. After five days of paid work for his client—mainly bridge inventory—Thoreau left their employment to pursue his own private investigation, becoming a silent third partner in the otherwise raucous public controversy.

To create the map Leslie pulled from the archive, Thoreau sailed, rowed, poled, and dragged his flat-bottomed boat up and down his three rivers like a voyageur, often assisted by his friend William Ellery Channing. Related tasks included a midwinter ice-skating reconnaissance of dangerous river openings, a midsummer investigation of water temperatures using a bottle-buffered thermometer, and a secret program of independently monitoring the opposition's hydraulic experiments. For eighteen months he used his cloth scroll map—and its extension for the lower Assabet—to sleuth out the entanglement of natural and cultural processes shaping the Concord River Valley.

Culminating Thoreau's project was his discovery of the river's *vitesse de régime,* its equilibrium size, shape, and channel roughness under steady-state conditions. He learned about this principle from a three-volume monograph pulled from the stacks of Harvard's library and translated into English: *Principes d'Hydraulique,* published by Pierre Louis Georges Du Buat, an engineer of the French Enlightenment. Ralph Waldo Emerson, perplexed to the point of amusement by his friend's obsession with river research, remarked in an August 3, 1859, letter to Elizabeth Hoar: "Henry T. occupies himself with the history of the river, measures it, weighs it, & strains it through a colander to all eternity."[7]

Why did Thoreau work so intensively on something so far removed from his literary, political, and ecological interests? Why did he remain invisible throughout the four-year legal process? What did he learn about river science and human impacts? How can his example help us manage today's pressing environmental issues? These were the questions that propelled me forward until I found the answers I share in this book.

THE BOATMAN BEGINS WITH A HISTORY of Concord's three rivers, with special attention to the human alterations shaping them before and after European colonization. It explores the centrality of these flowing streams to the life and work of Henry David Thoreau. It provides the first extended account of his behind-the-scenes work on the "flowage controversy." It publishes the results of Thoreau's rigorous river science for the first time. It proves that the joy he took in the outdoors was undiminished by his knowledge of how messed up it was: all of his three rivers had been enlarged, deepened, and locally invigorated by a spasm of agricultural and engineering impacts working their way through the system.

The introduction lays out four key ideas. Immediately after the 1854 publication of *Walden,* Thoreau found himself increasingly drawn to the Concord River watershed as the largest, wildest, and most beautiful thing in his daily life. There he found a coherent natural system where the power of

Homo sapiens—rather than of glacial ice or forest growth—had become the dominant geological agency. There he found a stream of time that mirrored the work schedules of upstream factories, a place where "the very fishes feel the influence (or want of influence) of mans [*sic*] religion," and where "all nature begins to work with new impetuosity on Monday." There he found scientific mysteries galore—the unexplored phenomena operating beneath his everyday observations of nature. Motivated by an insatiable curiosity, he plunged into the most rigorously analytical and quantitative work of his life.[8]

The first four chapters are of one piece. They merge the ancient geological story of the Concord River with the biography of its greatest admirer. Chapter 1 explores the archaeology and prehistory of the alluvial valley, a place the Native Americans called Musketaquid. This lazy river of grass, this "fertile and juicy place in nature," had previously been the flat, clay-rich bed of an old glacial lake. After lake drainage, slow tilting of the earth's crust kept the valley fertile enough to support a succession of human cultures. Chapter 2 narrates the concord and discord of America's oldest inland river town, established in 1635. Its historic archaeology of sawdust, musket balls, wrought iron nails, and broken English ceramics was superimposed on the prehistoric archaeology of the arrowheads, fishing weights, and clay pots of the previous geological epoch. Chapter 3 narrates the agroindustrial makeover of the valley between 1710, the time of its earliest gristmill, and the early 1840s, when the impacts of dams and bridges were reaching their peak and ramifying throughout the drainage network. Thoreau enters this chapter in the 1820s as a barefoot boy on the riverbank with dreams of voyaging the world via the scows of the Middlesex Canal. Chapter 4 opens with Thoreau as an angry young man threatening vigilante justice against the downstream Billerica dam. It ends with the covert, unethical sale of public water privileges to private industrial interests.

The middle chapters are also of one piece. They highlight Thoreau's life as a river boatman. Chapter 5 brings his biography to our side of his famous excursion to Walden Pond in the mid-1840s. It describes the boats he built, the saltwater voyages that inspired him, and the bustle of waterfront activity in his hometown. Chapter 6 opens with Thoreau's life-changing epiphany about wildness in 1856. He learned to see it not as something primitive and pristine, but as a force of nature, independent of human control, flowing continuously through his engineered landscape. He took full advantage of that wildness by sailing transient inland seas that were enlivened by human intervention. Chapter 7 describes the daily sojourning habits of the fully mature Thoreau, proves that he preferred the river to inland woods, and shares his river experiences with the reader through a representative year.

The last five chapters plunge Thoreau into the historical and intellectual gestalt of what would today be called environmental assessment. He enters and exits an egregious case study of river mismanagement in which honest labor met political "chicanery." Chapters 8, 9, and 10 narrate his deep dive into a series of forty-one discrete empirical and theoretical tasks between February 1859 and September 1860, an eighteen-month period of his life without precedent. These chapters interweave the vigor of Thoreau's scientific studies with the plodding legislative work of the General Court and the nuanced judicial decisions of the State Supreme Court. Chapters 11 and 12 bring Thoreau's biography and the flowage controversy to their untimely closings. As tuberculosis slowly claims the poet-scientist, the meadowland farmers win and lose their case through jurisprudence and corruption, respectively. The outbreak of the Civil War pushed this issue to the back burner of political priorities.

The conclusion offers a retrospective interpretation of Thoreau's pioneering insights about the "manners" of his rivers and explores their implications for modern environmental thought. For closure, the epilogue revisits the scroll map. The channels and meadows he mapped, though given up as a wet wasteland in 1862, have since been reclaimed as the Concord, Sudbury, and Assabet Units of the U.S. National Wildlife Refuge system, the largest patch of wild landscape in the greater Boston area. Twenty-nine miles of these watercourses were designated "Wild and Scenic Rivers" by an act of Congress in 1999, in part because of their historic and literary associations. Thoreau would have been pleased with the way things worked out.

MY ORIGINAL PLAN was to present Thoreau's quantitative data and his detailed descriptions as appendices to this book. As this project progressed, my editor, John Kulka, and I opted to create and maintain a specific online archive for all supplementary material. The URL for this online repository is http://robertthorson.clas.uconn.edu/writing/books/the-boatman/online-repository/.

Finally, this book presents Thoreau's story as a boatman and the changes to his rivers using historical methods and original source documents.

Fortunately, this version of the story can be independently verified and strengthened using geoarchaeological methods. I refer to sediment coring studies at strategic sites, field investigations of Anthropocene landforms, and mapping with LiDAR (light-detection and ranging) techniques. This work is in progress.

THE BOATMAN

Introduction

ON AUGUST 31, 1842, HENRY DAVID THOREAU DINED with the writer Nathaniel Hawthorne and his new bride, Sophia. The riverside setting was the Old Manse, where Ralph Waldo Emerson had written *Nature,* and where Henry had planted a lovely vegetable garden for the couple earlier that summer. After a dessert of the first "watermelon and muskmelon" from the garden, Hawthorne and Thoreau went out for a row on the Concord River, the first of many water sojourns they would share during their next four years of friendship.[1]

Hawthorne's memory of that experience, written the following day, provides the most compelling known description of Henry's skill as a boatman. He "managed the boat so perfectly, either with two paddles or with one, that it seemed instinct with his own will, and to require no physical effort to guide it." After their evening voyage, Thoreau impulsively offered to sell Hawthorne the boat for seven dollars. Hawthorne accepted, hoping that he could also "acquire the aquatic skill of the original owner." So on the following day Henry returned to the Old Manse to give Hawthorne a lesson, assuring him that "it was only necessary to will the boat to go in any particular direction and she would immediately take that course, as if imbued with the spirit of the steersman." This was certainly not the case when Hawthorne gave it a try. Under his inept command, the boat seemed "bewitched, and turned its head to every point of the compass except the right one." I personally understand Hawthorne's point, having capsized a leaky replica of this same boat when staging the frontispiece photo for this book. When Thoreau took the oars, however, it "immediately became as docile as a trained steed."[2]

THAT AUGUST HENRY WAS A TWENTY-SIX-YEAR-OLD bachelor and aspiring writer living in town with his family. For more than two decades the Concord River lay within a stone's throw or so of the family homes they lived in. Thus

FIGURE 3. Confluence in Thoreau's century. Herbert Wendell Gleason. Egg Rock, 1899. Westerly view showing the Sudbury River (left) joining the Assabet River (center) to become the Concord River, flowing north (right). Photo by Herbert Gleason (1899.12) on October 31, 1899. Courtesy of the Concord Free Public Library.

it was only natural that Thoreau developed a strong attachment to the river triumvirate in his backyard: the Sudbury, Assabet, and Concord. To the Puritan pioneers who settled this alluvial valley in 1636, these were the South, North, and Great Rivers, respectively. In map view, they look like three bent spokes converging on Egg Rock, the *axis mundi* of Thoreau's biography.[3]

Throughout the 1850s Thoreau considered this three river system to be "his own highway," the "only wild and unfenced part of the world hereabouts." This was because it could not be owned, and therefore tamed. Additionally, the sandy and loamy uplands of his hometown landscape had been more intensely deforested to create private farms. In contrast, the muddy, flood-prone, riparian forests of the bottomlands had been left largely intact or had become overgrown with willows, alders, and swamp maples. During the last decade of his life, Thoreau visited his rivers more than twice as often as the upland woods and lakes that he is far better known for writing about. Counterintuitively, the wildness of these channels was being intensified by a pulse of change being caused by deliberate land conversions over the surface area of the watershed, and by engineering projects at specific points along the lines of his streams.[4]

Despite Thoreau's lifelong preference for river scenery, his pine-scented, terrestrial, woodsy persona is affixed to his public image like a burr to a pair of trousers. Surely this is a response to his astonishingly successful masterpiece *Walden* (1854), originally published as *Walden, or Life in the Woods*. "I went to the woods because I wished to live deliberately," he famously wrote, forever branding our impression of him as a man of the woods first and foremost. During his manuscript revisions of *Walden*, Thoreau knowingly promulgated this distortion to mythify his remove from the society of his largely deforested river town. That's why his friend Bronson Alcott suggested he name the book *Sylvania,* which translates as "forest land."[5]

Consider Abigail Rorer's lovely cover painting for David Foster's *Thoreau's Country*. In the foreground is a farmer scything golden grain from a cultivated field likely underlain by dry sandy loam. In the middle distance are woodlots and pastures crisscrossed by fieldstone walls. And in the background is the cleared land of a barn-capped drumlin. Not one drop of water is visible. Foster's lovely text flows in the same general direction, uphill toward the pastures, woods, swamps, and interior ponds.[6]

The Boatman asks you, the reader, to turn around to see the wetter side of Thoreau country: The three blue highways of navigable water flanked by open bays, lush meadows, and rocky cliffs. A riparian ecosystem where the muskrat, rather than the woodchuck, is the celebrity rodent. A habitat where the otter, rather than the fox, is the most elusive predator. A willowy bank where Thoreau kept boats to sail inland seas, row rippled waters, and pole over submerged meadows where blooming flowers could be seen beneath clear water. Biographer Robert Richardson concluded that "Thoreau was as much a man of the rivers as a man of the woods." Boating companion William Ellery Channing went beyond that parity to claim that "the river [alone] was his great blessing in the landscape." Philosopher Alfred Tauber concluded: "the river was his poetic, if not existential, source of being."[7]

Indeed, no careful reader of Thoreau's journal can miss his lifelong enthusiasm for boats of all kinds and shapes. Adventurer that he was, he paddled away from Port Concord on thousands of days to access distant sojourning destinations during what he called "fluvial," "riparial," and "rivular walks." Shipwright that he was, Thoreau built at least three wooden boats in his lifetime, and rigged and retrofitted others. Writer that he was, he infused his literary craft with nautical language, whether from the Old World classics—Homer's epics, Norse sagas, Shakespeare's theater—or from his own boating experiences. Laborer that he was, he used his boat to harvest grapes, haul cranberries, collect driftwood, and clear snags. Scientist that he was, he employed his boat as a research vessel to fathom the shape of its channels and to study aquatic plants. Artist that he was, he imagined his boat as a floating studio, complete with oars sticking out the windows. Colloquially, Thoreau was the proverbial "river rat," foreshadowing the joy of Rat, Kenneth Grahame's character in *The Wind in the Willows:* "Believe me, my young friend, there is *nothing*—absolutely nothing—half so much worth doing as simply messing about in boats."[8]

"Walking" is probably Thoreau's best-known nature essay. Missing from his bibliography is a counterpart called "Boating," or "Rowing," or "Sailing," even though he had plenty of material to work with. "No wonder men love to be sailors," he penned after an invigorating day's sail over Concord's Great Meadow, "to be blown about the world sitting at the helm, to shave the capes and see the islands disappear under their sterns. . . . It disposes [one] to contemplation, and is to me instead of smoking." Indeed, sailing was both a stress reducer for Thoreau and a lifelong addiction. Boating gave him the "searoom" he needed from society, and the freedom to go wherever the "current" of his thoughts sent him. He advises us to keep moving, and never to be caught "lying on our oars." He had about him the "suggestion of a seafaring race," wrote Thoreau's young friend Edward Emerson, Waldo's son, who admired Henry's unbridled enthusiasm for his water world.[9]

Thoreau's three rivers were almost daily pathways to higher country, even when he left his boat tethered to a tree. "I can easily walk 10 15 20 any number of miles commencing at my own door without going by any house— without crossing a road except where the fox & the mink do. . . . First along by the river & then the brook & then the meadow & the wood-side—such solitude." This sequence follows the drainage network up from the main stem to tributary, spring-fed source, and finally upland forest. More often than not, Thoreau's ticket to the upland woods and lakes was a flowing stream. And, as Channing once said, "a trip up the river rarely ended with the water, but the shore was sought for some special purpose."[10]

Another lopsided result of Thoreau's posthumous literary success with *Walden* is our tendency to affix his personal image to its isolated namesake pond in the woods rather than to the sprawling waterways that so dominated his life. Yes, Walden Pond was the perfect place for a young idealist preoccupied by the cyclical rebirth described in Oriental scripture and keenly interested in his own transcendental reform experiment. A lake so deep that some thought it bottomless. So clear it was a lower heaven unto itself. So isolated it was a miniature cosmos. A glacial kettle so steep-walled that the focus was inward and downward. So pure that neither weeds nor muskrats claimed its stony shore. A place of loons rather than of gulls. So timeless that its history faded to eternity. Walden Pond remains a stunning place, aesthetically, historically, and spiritually, which explains why his house site there has been a place of international pilgrimage for more than a century.

But we must not forget that Henry's excursion to Walden was a time apart from the everyday routine of his family life, a place apart from the settled village he lived in, and a terrain apart from the moist, flat meadowlands of his Main Street life. When living near the stony, droughty shore of Walden's ascetic water, Thoreau was simultaneously enjoying the muddy fertile meadows of the nearby rivers. Looking back from the late 1850s, and with *Walden* a distant memory, he confided to himself:

> I was born upon thy banks, River,
> My blood flows in thy stream,
> And thou meanderest forever
> At the bottom of my dream.[11]

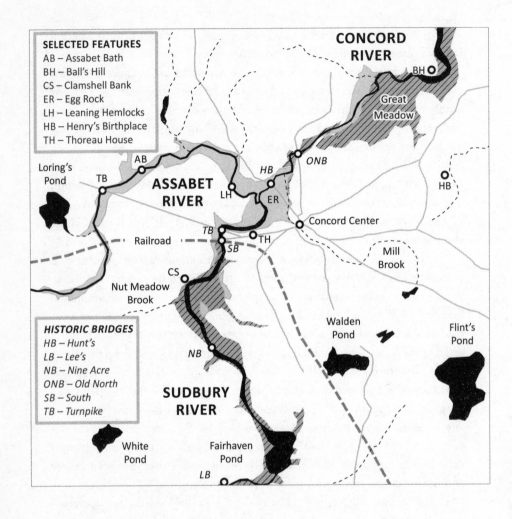

SELECTED FEATURES
AB – Assabet Bath
BH – Ball's Hill
CS – Clamshell Bank
ER – Egg Rock
LH – Leaning Hemlocks
HB – Henry's Birthplace
TH – Thoreau House

CONCORD RIVER

BH

Great Meadow

Loring's Pond

AB

TB

ASSABET RIVER

LH

HB

ER

ONB

HB

Concord Center

Railroad

TB

SB

TH

Mill Brook

CS

Nut Meadow Brook

HISTORIC BRIDGES
HB – Hunt's
LB – Lee's
NB – Nine Acre
ONB – Old North
SB – South
TB – Turnpike

NB

SUDBURY RIVER

Walden Pond

Flint's Pond

White Pond

Fairhaven Pond

LB

FIGURE 4. Thoreau's three rivers. Map showing selected places and features near the confluence. Meadows (hachured) and sandy alluvium (gray) mapped by U.S. Geological Survey. (Stone and Stone 2006). Abbreviations for bridges are in italics.

Yes, he built a small house at Walden Pond, slept there, wrote there, and hoed beans nearby. Yes, it was his defining, transformative intellectual experience. And yes, it will properly remain the single place we most clearly associate with this American genius. But it was never his home, his instinctual "sense of place." That was the greater Concord River, settled by both Native Americans and by Puritan settlers at the triple point where its three main streams coalesce. Henry's greater affinity for the river was self-evident to his personal friends and first biographers—Ellery Channing, Franklin Sanborn, and Ralph Waldo Emerson—who said so matter-of-factly. In his funeral eulogy, Emerson described a "river on whose banks" Thoreau was "born and died." In contrast, the woodlot at Walden Pond was the locus for a two-year experiment in deliberate living, and its one-room house the site of a writing retreat beyond the reach of Emerson's shadow and quieter than the crowded Texas House, where his upwardly mobile family had just moved. Appropriately, Channing called Thoreau's small Walden house a "wooden inkstand."[12]

Thoreau's lifelong attachment to his three rivers explains why *Walden, or Life in the Woods* is bookended with river experiences. Its epigraph—"I would crow like a chanticleer in the morning, with all the lustiness that the new day imparts"—was conceived during a boating trip in the fog on both the Sudbury and Assabet Rivers. One of *Walden's* final conclusions, "The life in us is like the water in the river," obviously doesn't refer to a stationary pond where the water is exchanged only once every five years. One of *Walden's* chapters, "Baker Farm," is named for a piece of riverfront property that Thoreau contemplated as an alternative site for his experiment in social reform. "Dwell as near as possible to the channel in which your life flows," he wrote in a book about a "deep green well" with neither inlet nor outlet channels. "Time is but the stream I go a-fishing in," he imported to a book about timeless eternity. In short, much of Thoreau's masterpiece was distilled from nearby river experiences.[13]

Spiritually, the waters of *Walden* evoked purity and stillness. In contrast, those of his rivers evoked richness and flow. "Man must not drink of the running streams, the living waters, who is not prepared to have all nature reborn in him." In watershed brooks, "a man's life is reborn with every rain." Boating his "silver-plated stream" was "like embarking on a train of thought" that "meandered through retired and fertile meadows far from towns . . . to lurk in crystalline thought like the trout under verdurous banks." His rivers carried "the blood of the earth." They were its "blue arteries pulsing with new life."[14]

The only other book Thoreau published in his lifetime was *A Week on the Concord and Merrimack Rivers*. Its inspiration was an 1839 vacation trip from teaching: a camping excursion to the White Mountains with his older brother, John Jr. To get there, they boated down the lower Concord, through the Middlesex Canal, and then up the Merrimack to the limit of navigation at Hooksett. Following John's tragic death from lockjaw two years later, Henry decided to memorialize his brother with a book-length account of the time they spent together, even though John is never mentioned by name. The result was Thoreau's "big book," *A Week*, written as two complete drafts from his desk at Walden between 1846 and 1847, and published two years later, in 1849. Since then, *A Week* is usually considered Thoreau's main literary account of the Concord River.

This distorts reality. For starters, the Thoreau brothers spent less than one day of their fourteen-day excursion on the lowermost, canal-like reaches of the Concord River. They passed most of their time either on the larger, south-flowing Merrimack River or hiking the high country. And though the title *A Week* suggests a travel narrative, this aspect of the book was mainly a literary device to move the reader through an otherwise disjointed text. The bulk of the book is a scattered anthology of transcendental musings, "doggerel" verse, natural history description, and events unrelated to the fictional chronology, which distills two weeks into one. After an excellent introduction drawn from much later experiences, *A Week* quickly becomes for many readers the least favorite of Thoreau's works, difficult to get through even as a research task. This explains why, within his lifetime, it was the archetype of a self-published commercial failure. Hundreds of unsold copies gathered dust in his garret to the end of his life.[15]

Henry's unheralded river book is his journal. We're talking forty-seven manuscript volumes of handwritten pages containing more than two million words written over twenty-four years. Increasingly, scholars consider this to be his chief literary work. And the chief focus of that work is the river. Beginning in mid-1851, regular daily journal entries contain thousands of astonishing observations and philosophical reflections linked to flowing water in some way. Those rivers—not Walden Pond—were the main channels of Henry's mature life. Physically as a traveler in Concord and metaphorically as a writer. Symbolically, his reference books for his journal were kept on bookshelves built of river driftwood "brought down by the spring freshets." Before holding his words, this wood had been "the perch of turtles and the dining-table of clam-loving muskrats."[16]

Walden Pond and Thoreau's river country need not compete with each other for our attention. Rather, they inspired complementary literary works: his masterpiece *Walden* and his life's work, the *Journal*. Pond and river are complementary geometric forms: a circle and line, respectively. They are also complementary geographic icons for the three main epochs of Thoreau's full life. First and last were the river. Walden occupied the center.[17]

"MY WINDOW LOOKS WEST," Thoreau wrote from the high overlook of his attic sanctum in the Main Street house. The river was about five hundred feet away: comparable in distance to the view between his house at Walden Pond and the main part of the lake. As within his Walden house, Thoreau arranged his bed so that he could see the distant water first thing in the morning. "Before I rise from my couch, I see the ambrosial fog stretched over the river, draping the trees . . . as distinct as a pillow's edge, about the height of my house." The frame of this window put the edge of the watershed on the horizon, the neighboring houses in the foreground, and the Sudbury River crossing the focal point from left to right, in this case "from southwest to northeast." Views through this frame—of mists, sunsets, reflections, colors, ice, plant growth, autumnal tints, and moving boats—were like time-lapse photographs taken from a stationary camera. Conversely, this window was a homecoming beacon for Thoreau when he was out on the water: "Rounding the Island just after sunset, I see not only the houses nearest the river but our own reflected in the river."[18]

Physically, the river lay across the street from the family home on the far side of Channing's house lot. Below the muddy, willowy bank was his "boat place" where he kept his vessel partially hidden between March breakup and December freeze-up. This "harbor" bisected the lowermost broad bend of the Sudbury River, which roughly parallels west Main Street. It lay about two thousand feet above the confluence, marked by Egg Rock, a glacially smoothed knob of diorite. This launching place offered Thoreau a daily menu of three river entrees. "I have three great highways raying out from one centre, which is near my door," he wrote during the outset of his river project. He could travel

FIGURE 5. Western window. West side of Thoreau-Alcott House at 255 Main Street, Concord, taken at sunset to highlight the windows of Henry's third-floor sanctum. The one-story addition in foreground was a later addition. August 2016.

"down the main river or up either of its two branches. Could any avenues be contrived more convenient?" One year earlier he exaggerated: "I think that I speak impartially when I say that I have never met with a stream so suitable for boating and botanizing as the Concord, and fortunately nobody knows it."[19]

Deciding which of his three water highways to take depended on the combination of his mood and the direction and strength of the wind. "Up Assabet" is probably the most common opening phrase in Thoreau's two-million-word journal. This was his favorite destination under default conditions, meaning the wind was light and the river stage was neither in flood nor in drought. His default experience was to row upstream for several miles and then drift gently downstream under the influence of the current. Franklin

Sanborn, Thoreau's friend and occasional boating companion, described it as the best place in town "for rowing and retirement." Indeed, there were no bridges to pass beneath, from which someone might accuse Henry of idling his life away. The Assabet stayed open longer in winter, thanks to its stronger current. The three-mile reach below its Union Turnpike bridge was a "picturesque" meandering stream with sandy cut-banks alternating with beach-like sand bars and swimming holes. One favorite destination was the Leaning Hemlocks. Being nestled in the deep shadows below Nashawtuc Hill to the south, and being watered by cool moist seepages, the riverside flora at the hemlocks was wilder and more boreal in its aspects than elsewhere along the river.[20]

"Upriver," "To Fairhaven," and "To Clamshell" were other common opening phrases. This signified a southward trip up the lakelike Sudbury River, which zigzagged between rocky cliffs and lush meadows. Its current could be detected only at narrows and shallows. Parallel lines of sweet-scented lily pads, aquatic flowers rooted in shoreline mud, and buttonbush swamps flanked the river for much of its length. With a breeze from the north or east, he would usually sail. Otherwise he would row. In winter he skated a series of linear rinks. Clamshell Bank was the epicenter of his archaeological universe. Further upriver was Fairhaven Pond, his most frequent southerly destination, where he could go sailing even during times of low water. There he found "the wildest scenery imaginable,—a Lake of the Woods," which he described as comparable to those he'd seen in Maine, with "forests and the lakes and rivers and the mountains." When he was away from town, it was Fairhaven that he missed most.[21]

"Downriver," "To Great Meadow," and "To Hill" signified a northward trip down the Concord River below the triple point of the confluence. After passing through a straight reach aligned by the local bedrock strike, arched by two bridges, and flanked by gravel bars of historic sediment that were repeatedly dredged, he entered the north side of Great Meadow. Bounded by the site of the Old North Bridge to the southwest and Ball's Hill to the northeast, it was two miles long and half a mile across. When in flood, the meadow was his favorite inland sea to sail upon because the wind was least impeded and the waves were highest. In winter, it became a quasi-arctic scene of enormous snowdrifts. Below Ball's Hill, the river widened into something resembling one of the Finger Lakes of New York, and for much the same glacial reason. Far downstream and west of Billerica were "water prospects of a larger river"

FIGURE 6.   Boat place. South side of Sudbury River below Main Street just east of Nashawcut Bridge shows character of modern shore in vicinity of Thoreau's boat place. August 2016.

reflecting a "grand range of hills, somewhat cliffy" creating "one of the most interesting and novel features in the river scenery." Left unsaid is that these reflections came from a reach made broader and more mirrorlike by a downstream factory dam.[22]

Anthropocene Insight

On the sparkling morning of August 11, 1859, Henry was rowing the lower Assabet, making measurements, gathering samples, sketching maps, and taking it all in. Drifting toward him on the smooth current was a parade of iridescent clamshells "floating down in mid-stream—nicely poised on the water," each "with its concave side uppermost." Each was a "pearly skiff set afloat by the industrious millers." His delight was one more tidbit of support for a big idea he'd been working on all summer: that he'd become "part and parcel" of nature in a brand-new way. Not in the romantic, holistic, transcendental, and Arcadian way of his intellectual youth, when under the mentorship of Ralph Waldo Emerson, and not in the organic, evolutionary way of his late-career botanical studies, when following the anatomy and taxonomy of Asa Gray, but in a new and fundamentally historical way that didn't yet have a name. Today we would call it his Anthropocene insight: the recognition that his human agency was so completely interwoven with nature that his geological epoch was unique in universal history.[23]

During the previous week, Thoreau had been monitoring the rise and fall of the water-surface elevation, a parameter called "stage" by hydrologists. By mid-August he had gathered enough high-resolution data to explain why these heavily mineralized shells could float downstream without tipping and sinking. He compared each to an "iron pot" made to float by setting it down gently on still water. Being curious, he touched one of the shells with an oar, and watched it "at once sink to the bottom." Others sank with the slightest riffle or breeze. That's when he realized that this flotilla of precariously balanced clamshells was a highly unlikely yet fascinating phenomenon arising from the tangle of interactions between his species, clams, and muskrats.[24]

In general terms, progressive deforestation by agricultural humans enhanced storm runoff, thereby raising flood discharges. This was increasing the power of brooks to acquire and transport mineral sediment. This improved the habitat of freshwater clams (mussels) downstream from bends by concentrating the fine-textured mineral mud they prefer. Meanwhile, muskrat habitats were being improved by wetter summer conditions associated with dam construction and cleared watersheds, and by the booming guns and snapping traps that were killing off their predators. Finally, industrious humans were building milldams across the brooks with gates that could be closed to hold back the flow and opened to turn water wheels. All three species— *Homo sapiens, Unio fluviatilis,* and *Ondatra zibethicus*—were components of a single integrated system responding to the human makeover but not orchestrated by it.[25]

Specifically, "during the night on the shore" of August 10, 1859, nocturnal muskrats dug clams from sediment pollution, leaving the remains of their meals concave side up just above the waterline. During the "forenoon" of August 11, diurnal humans went to work and opened the dam gates, creating a gently rising tide of water. The empty, inverted shells were gently lifted "afloat," pulled to the center of the channel by its slightly faster flow, and carried downriver toward Thoreau's boat. Prior to dam construction, this phenomenon would have been rare on the Assabet because its river stage usually falls continuously for many days before rising abruptly and turbulently during rainstorms. And prior to gathering his quantitative stage data, Thoreau would not have realized the significance.[26]

Welcome to Thoreau's new world. For better and worse, human activity was now the dominant agency driving landscape change. The dominant power remained solar radiation, which created warmth, wind, precipitation, and organic growth, as before, and which powered human existence as well. But the manifestations of that power were being nudged and coerced into new directions by Euro-American culture and technology. This was true not only for obviously engineered places, such as roads, villages, farms, bridges, canals, railroads, and cleared fields, but also ubiquitously in the background.

Thoreau's appreciation of the pervasiveness of human intervention wasn't a moral judgment about whether humans had the God-given right to remake the landscape to their specifications, though most of his neighbors would agree that they did. Thoreau disdained this sort of thinking as harmful to human spiritual growth because it contributed to the "egotism of the race," a vanity that culminated in "man-worship." Rather, his insight was the straightforward, cognitive recognition that the boundary between nature and non-nature had disappeared. Humans had become "of the Earth earthy" in a way very different than at their hominid origins. "Nature looking into nature" now meant something new. Borrowing the language of environmental historian Michael Rawson, Thoreau's "ideas about nature, as ethereal as they might seem," gave him "important frames for structuring social thought."[27]

SEVEN YEARS EARLIER, IN 1852, when Thoreau was rescuing his dormant *Walden* manuscript, he held rather conventional ideas about the difference between wild and tamed landscapes. At that time he was still struggling with the biblical baggage brought to this continent by the religious separatists who had landed at Plymouth Rock and by the devout Puritans who had followed, the "People of the Book." He was still trapped by the cultural paradox that historian William Cronon called "the trouble with wilderness" in his influential 1996 essay of that title. The paradox is that "wilderness embodies a dualistic vision in which the human is entirely outside the natural. . . . The place where we are is the place where nature is not." In short, wilderness is not an actual place but an imagined place, a cultural construct, a flight from history, and a tautology.[28]

But by August 1856, the revisions, publication, launch, and reviews of *Walden* were behind him. That's when Thoreau shed his trouble with wilderness with an extended epiphany culminating with the statement: "It is in vain to dream of a wildness distant from ourselves. There is none such." Thereafter, he abolished the paradoxical dualism that was still confusing Ralph Waldo Emerson with his "me" of human consciousness and the "not me" of everything else. For the pioneering conservationist George Perkins Marsh, the dualism was titled *Man and Nature*, a bipolar vision with Christian stewards at one pole and everything else at the other. In the twentieth century, Leo Marx invoked the "machine in the garden" as a modern stand-in for the dualism of the Tree of Knowledge within the Garden of Eden. In this century, dualism still rears its head for historians as Richard White's technology versus nature, David Nye's first versus second creation, Richard Judd's first versus second nature, and Diane Muir's economy versus ecology. The only real dichotomy is the taxonomic human versus nonhuman. This, however, is biologically no more significant than lobster versus non-lobster.[29]

This post-*Walden* Thoreau would have taken serious issue with Bill McKibben's *The End of Nature,* which argues, somewhat apocalyptically, that nature is dead because humans killed it. Though it's true that no place in nature remains separate from human influence, it does not follow that nature is dead. Since the very beginning of geology as an organized discipline in 1807 in London, it's been clear that earth history is a book of discrete chapters, epochs in which one operating system—paleogeography, paleoclimatology, paleontology—replaces another. Nature did not end when bacteria invented photosynthesis 2.5 billion years ago, even though the entire globe became polluted by their exhaust gas (oxygen), which led to the creation of nucleated cells and eventually to a burst of evolution leading to us. Nor did Nature end when our upright ape ancestors invented planet-changing technologies leading to the controlled use of fire, stone tools, animal domestication, horticulture, the steam engine, and nuclear fission. In the *longue durée* of geology, the rise and dominance of our eusocial species is no big deal. Paleontologist Stephen Jay Gould put it this way: "Our planet is not fragile at its own time scale, and we, pitiful latecomers in the last microsecond of our planetary year, are stewards of nothing in the long run."[30]

Consider the alluvial muck that brought settlers to Concord. It's a young geological stratum of variable thickness that is no older than about 14,000 years at its base and is modern at the surface. The deepest and oldest stone flakes and projectile points of ancient peoples within it define a local stratigraphic marker for the approximate base of the Holocene Epoch (the last interglacial), which succeeded the Pleistocene Epoch (the last ice age). Similarly, the deepest and oldest sawdust within it is a stratigraphic marker for the onset of a new—still unofficial—geological epoch, the Anthropocene, which replaced interglacial business as usual. Each of these three epochs has its own characteristic flora and fauna. From Pleistocene to Holocene, the fauna went from mammoth, mastodon, sloth, bison, and New World horses to moose, caribou, deer, bear, cougars, wolves, and beavers. From Holocene to Anthropocene, it went from these "nobler animals" (Thoreau's phrase) to the imported invasive species of Old World livestock (cattle, hogs, sheep, horses, poultry) and to indigenous critters best suited to disturbed land. Thoreau understood that the next faunal turnover was only a matter of time, that his "Saxon race" and its familiar animals "would become extinct" and "disappear from the face of the earth." He asks us: "Is not the world forever beginning and coming to an end, both to men and races?" To the geologists out there, the answer is always yes.[31]

Thoreau's grasp of the Anthropocene as a concept—though not by that name—did not conflict with his interest in wildness, which remained a very useful idea. He believed that "wild" came from "willed" and was thus a useful umbrella term for any phenomenon or attribute beyond human control, either because we don't want to exercise control (in the case of backyard birds) or because we can't (in the case of Icelandic volcanic eruptions). But there's plenty of wildness out there that has nothing to do with humanity or willfulness. I refer to the ubiquitous and irrepressible "force of nature" that is much more complex than the fundamental physics of gravity or magnetism. It's spontaneous and unpredictable—for example, the exact place and time an acorn will drop into Thoreau's boat. Mathematically, it's probabilistic, as opposed to deterministic. At higher levels such wildness includes the spontaneous emergence of self-organized, nonlinear systems such as river ripples, potholes, and meanders.[32]

Long before the rise of complexity theory and chaos studies in the 1980s, river geologists were studying what they called the "complex response." They discovered that the general outcomes of identical watershed experiments always turned out the same, but the specific outcomes were always different. The basic idea is that in any cascade of causation, outcomes are always contingent on previous outcomes. With dominos being tipped over, this contingency is fairly simple. But with a turbulent fluid in a natural river channel, the specific consequences are always unpredictable, chaotic, and therefore wild to some degree. In Thoreau's day, the construction of the Union Turnpike Bridge over the lower Assabet River about 1827 perturbed the river, which responded with new currents, banks, beds, curves, clams, muskrats, landslides, and leaning trees. Writer Wallace Stegner, a twentieth-century devotee of Thoreau, summed it up this way: "Order is indeed the dream of man, but chaos, which is only another word for dumb, blind, witless-chance, is still the law of nature."[33]

Thoreau saw the perturbing turnpike bridge as a "pretty picture" worthy of a detailed sketch and long technical description. On the first sunny morning after several days of rain, the river was running strongly. Standing on a sandbar below the bridge and looking through its single arch, he saw it "partially damming the stream." Upstream the water was several feet higher, smooth, and reflective. Downstream it rushed to a powerful "turbulent" fall with a foaming, symmetrical double curve. Aside from the fluids of air and water, everything about this scene was a consequence of human action: the quarry from which the stone was hewn, the bridge built of those stones, the second-growth trees rising in the woods, the unplanned double waterfall, and even the sandbar he was standing on, which consisted of sediment eroded from somewhere else. To envision this as an apocalyptic transformation caused by human beings, or the residue thereof, would have ruined his day, and all the rest to come.[34]

So he chose to reframe his idea of wildness. No longer would he see his landscape as something being crushed by the nineteenth-century makeover he was contributing to as a surveyor. Historians Jo Guldi and Richard Armitage date this apocalyptic strain of thought as beginning about this time. Richard Judd calls it a "narrative of declension and destruction" typified by Rachel Carson's 1962 book *Silent Spring*. Instead, Thoreau saw the one-arch bridge as the precipitating perturbation in a cascade of physical wildness that would propagate many miles downstream and upstream in what legal scholar Jed Purdy calls "uncanny" ways. Realizing that his landscape was being underwritten by unseen causes, Thoreau began a program to investigate and explain the emerging appearances of his daily backyard life. This led him to a powerful insight involving a then-famous place on the Assabet known as the Leaning Hemlocks (see Figure 21).[35]

Fifteen years earlier, as a young naturalist, Thoreau had boated there with Nathaniel Hawthorne, taking great delight in dynamism. Hawthorne, in *Mosses from an Old Manse,* described the scene in 1846: "There is a lofty bank on the slope of which grow some hemlocks, declining across the stream with outstretched arms as if resolute to take the plunge. . . . A more lovely stream than this has never flowed on earth." His literary association would, within the next half century, cause the Leaning Hemlocks to become one of the most frequented picnic and excursion sites for Victorian Concord. In his *Concord Guidebook,* George Bradford Bartlett wrote that "every poet, philosopher and storyteller of Concord has delighted to sing the praise" of the hemlocks.[36]

In contrast, Thoreau was quite clinical in his 1852 journal description: "There are many larger hemlocks covering the steep side-hill forming the bank of the Assabet, where they are successively undermined by the water, and they lean at every angle over the water. Some are almost horizontally directed, and almost every year one falls in and is washed away. This place is known as the 'Leaning Hemlocks.'" Seven years later, during his 1859 river project, Thoreau discovered that the trees were then being tipped backwards, instead of forwards. He properly interpreted that they were on a slump or "slide" of land caused by even higher rates of bank undercutting being driven by a spasm of change precipitated by human impacts. The message for us is that—given the right attitude—one can find beauty and wildness in even the most devastated corners of nature.[37]

A more biological version of Thoreau's mental reframing regarding wildness involved the diversity of songbirds and small animals that were his daily companions and which fill the pages of his journal. "His Most serene Birdship!" was the bluebird, whose early migration coincided with snowmelt; "his soft warble melts in the ear." This was one of many migrating species that appeared in an impromptu list from early May: "Robins, song sparrows, chipbirds, bluebirds, etc., I walked through larks, pewees, pigeon woodpeckers, chickadee *tull-a-lulls*, to towhees, huckleberry-birds, wood thrushes, brown thrasher, jay, catbird, etc., etc. . . . Hear the first partridge drum. The first oven-bird . . . Blackbirds are seen going over the woods with a chattering bound to some meadow . . . creeper . . . warblers . . . myrtle birds . . . purple finch . . . bank swallow . . . vireo . . . yellowbirds . . . blackbirds . . . crows . . . hummingbirds."[38]

Many, if not most, of these birds would not be there had the forests not been cleared and the land not been managed in patches. They require either the ecotones of forest edges or the vigor of juvenile woodlands, what Thoreau called "sproutland." "Raccoons, skunks, rabbits, bobcats, bear, and lynx" also prefer this kind of disturbed habitat. All but the bear were celebrated in Thoreau's journal of local sojourns. All were drawn to the river, where the nutrient flux was highest in this otherwise sandy, gravelly, rocky country. These animal populations were thus wild responses to human interventions. Again: "It is in vain to dream of a wildness distant from ourselves."[39]

My main message here is that fans of Thoreau's lyrical nature writing must know that he was writing about a highly disrupted Anthropocene landscape rather than a quasi-stable Holocene one, not only in ecology and the abundance of huckleberries, but also in hydrology and geomorphology. "How many aspects the river wears," he wrote, "depending on the height of the water, the season of the year and state of vegetation, the wind, the position of the sun and condition of the heavens, etc., etc.!" To this list he could have added "the degree of human influence, etc., etc.," for by then it was as pervasive as the weather.[40]

The human "aspects" Thoreau enjoyed ranged from the violent breakup of river ice during the spring freshet to the "unctuous" stagnancy of the water during late-summer drought. Both scenes were influenced by deforestation and impoundment. Some of his sojourns were skinny-dipping fluvial walks taken during the suffocating heat of summer, which says something about how private his river refuges had become in a largely deforested country. Others were sprints of speed skating on the windswept river, during which he used his coattails as parasails. Land clearing and dam construction had improved the skating by chilling and stilling the water, respectively. Thoreau's daily river adventures were populated with fauna adapted to the human presence—for example, basking tortoises (occasionally the object of thrown snowballs) and red foxes (easily tracked during the fresh snowfalls of freeze-up). All of these joyful episodes were contingent on the human makeover in some way. Knowing this did not diminish his appreciation for the beauty or the excitement of his river life. Thoreau's positive attitude can help us brace for the global changes heading our way.[41]

Worlds in Collision

Thoreau's emerging views of the human makeover in the Concord River Valley culminated with the "flowage controversy," a protracted legal fight that ground its way through the courtrooms of Boston and Concord between 1858 and 1862. On one side was a gathering of Middlesex farmers from seven towns—Carlisle, Bedford, Concord, Lincoln, Sudbury, Wayland, and Weston—who had petitioned the state legislature for the removal of a downstream dam. At stake was a tract of organic alluvial soil so vast that it covered "ten to fifteen thousand acres" and was so rich that it drew Puritan settlers into the wilderness in 1636 to found the first colonial town above tidewater. Hay cut from those meadows fed the cattle that provided the manure that fertilized the fields that gave English Puritans the grain that gave them their daily bread and the milk they sopped up with it. Four generations later the descendants of "every Middlesex village and farm" dropped their tools to arm themselves against British imperial rule. Four generations after that first "Concord Fight," their descendants were once again rising up against tyranny, in this case against downstream industrialists.[42]

The dam in question was raised between 1794 and 1798 to plumb the Middlesex Canal, a state-chartered transportation corridor connecting Boston and the future industrial city of Lowell. That's when a low, leaky, timber-faced earthen dam built in 1710 for a colonial gristmill was raised two feet and strengthened to hold enough water to float commercial barge traffic. Thirty years later, during the heyday of the canal in 1828, this dam was raised

FIGURE 7. Billerica dam. Ice-covered reservoir of the Concord River flowing over Billerica dam and falling eight feet into splash pool. Sharp horizontal line above snow-covered rocks is the crest of dam. View is upriver, easterly. January 2015.

another foot, rebuilt even more strongly with cut blocks of granite, and lifted even higher during summer with wooden flashboards. By the mid-1840s, however, the canal was headed toward bankruptcy, thanks to the efficacy of railroads. Salvaging what profit they could, the private shareholders of the Canal Corporation sold their water "privileges" at bargain basement prices to downstream industrialists, Messrs. Charles and James Talbot at one textile factory and Mr. Richard Faulkner at another. The buildings their factories occupied are still there, bearing their names.

Upstream farmers argued that the Billerica dam was ruining their livelihoods by preventing proper drainage of the natural meadows during the summer haying season, and later during the cranberry harvest. Flotsam composed of rotting black hay and fermenting red cranberries marked the strandlines of more frequent and higher late summer floods. These were symptoms of a national transformation being driven by the courts that was particularly acute in New England. "By the middle of the nineteenth century," wrote legal scholar Morton Horwitz, "the legal system had been reshaped to the advantage of men of commerce and industry at the expense of farmers, workers, consumers, and other less powerful groups within the society." As a consequence, concluded historian John Cumbler, "rural agricultural society was giving way to an urban industrial society." Beginning with the Mill Acts of 1795, New England judges increasingly supported the theory that industrialization conferred the "greatest good for the greatest number," thereby creating "the world of the future, of cotton mills replacing saw mills and textile workers replacing farmers." Not coincidentally, 1859 was the very first year that the value of "U.S. industrial products exceeded that of agricultural products."[43]

As legal petitioners, the meadowland farmers asked Henry Thoreau to help paint the Billerica dam as a great evil. They paid him to boat the full length of Musketaquid and to interview old-timers, activities he loved. As legal respondents, downstream industrialists argued that the chronic wetness was due to upstream causes, that removal of the dam would cause hundreds of citizens to lose their jobs, and that their water privileges had been guaranteed by state law. They hired James B. Francis—a Goliath of American hydraulic engineering—to be their technical consultant and expert witness.

Francis testified with great authority for the factory owners. Henry never testified for the farmers. Not once is he mentioned in the 607-page, hardcover *Report of the Joint Special Committee upon the Subject of the Flowage of Meadows on Concord and Sudbury Rivers*. Based on this 1860 report, the

House and Senate of the Commonwealth of Massachusetts passed an act to tear down the dam, legislation strong enough to survive a legal challenge in the state's Supreme Judicial Court. Sadly, this massive legal bulwark was later repealed in 1862 by vested financial interests in a case that stinks of political corruption.[44]

Though Thoreau is never mentioned in the court proceedings, we know he closely followed the case to its bitter end, if only for intellectual amusement. His private journal paraphrases oral testimony given under oath, suggesting he attended the hearings. The people he interviewed became witnesses. His client shared his ideas in court anonymously. His private papers contain published pamphlets about the controversy, and handwritten copies of reports. His work schedule was timed to coincide with key dates of experiments and hearings. One of the movement's leaders, David Heard, was Thoreau's client, correspondent, and assistant. John Shepard Keyes, Concord's most powerful politician, personally gifted Henry with a beautiful blue-marbled copy of the full report, inscribed to "H. D. Thoreau, Esq. with regards of J.S.K." Simon Brown, chair of the River Meadow Association and former lieutenant governor of Massachusetts, thanked Henry with carriage rides around town after he became too ill to sojourn.[45]

The attorneys for opposite sides used lofty rhetoric to spin the situation in their favor. Presenting the case for the petitioners, the golden-tongued Henry French intoned: "This rich and beautiful expanse, forming . . . a noble feature of a delightful landscape, is converted into a loathsome laboratory of mephitic gasses and poisonous exhalations." "Instead of 'the sweet and wholesome odor of the new-mown hay,'" there is "'a foul and pestilent congregation of vapors' . . . as perceptible as the effluvia of a slaughter-house. Whole families have been prostrated and decimated by fevers. . . . Chronic and acute rheumatisms have become alarmingly prevalent. . . . Books and clothing contract moisture and mould in closets, trunks, and drawers. . . . The sale of real estate is paralyzed. . . . Some of the most substantial and pleasant dwellings have been long vacant, and . . . have found no purchasers. Strangers who come to view these estates, say, 'You have too much water here,' and go their way."[46]

Presenting the case for the respondents, the silver-tongued Josiah Abbott claimed on behalf of the industrialists that the whole case was a "great and magnificent mistake," a sentimental fiction derived from youthful nostalgia for the romanticized good old days. It was a "monstrous exaggeration," he asserted, to suggest that a "most beautiful and smiling mead" had been converted into a "barren wilderness" by the actions of one small dam, or that "an almost earthly paradise, whose beauties should be embalmed in song, and so go down to remote generations," had been converted "to an unsightly bog and an unhealthy quagmire."[47]

As this rhetoric ricocheted back and forth, Thoreau was busy proving to himself that both sides were correct, and that both sides were ignorant of the complexities. The downstream dam was indeed a serious problem for meadow agriculture. Simultaneously, the upstream watershed was being transformed in ways that made the lowland alluvial valley of Musketaquid wetter, especially during summer. But within the legal arena, the whole case—four years, four legislative acts, and great cost—boiled down to the single discrete question: was the downstream dam the *main* problem? No one knew because no one could have known; the science wasn't there yet, at least not in the public domain. Round and round the blame game went, creating a gold mine for the lawyers, who thrived in the vacuum of scientific knowledge and theory.

Two worlds were in collision. The view looking upstream was that of Lowell, Massachusetts, the hydropowered heart of Yankee industry. That's where the powerful Merrimack River, having drained from New Hampshire's highest mountains, fishhooks to the northeast. That's where the controlled fall of water turned 31 mills, 6,300 looms, and 225,000 thread spindles in that city alone. Farther down, the Merrimack flowed to the Atlantic as "mere *waste water*," to use Thoreau's term. More cloth was woven in this vicinity than in the entire Confederacy combined. Cotton picked by enslaved human beings was being manufactured into fabrics by America's most enslaved river. Equally enslaved was the lowermost reach of the Concord River, which enters the Merrimack at Lowell. There, above the locks of the Middlesex Canal, were four factory dams. For these industrialists looking upstream, the entire alluvial valley of Musketaquid was one large reservoir of gravitational potential energy created to power their factories before the age of coal-fired steam. For farmers looking downstream, what hung in the balance was their way of life, which they themselves had been slowly transforming during the previous three or four decades.[48]

ON AUGUST 23, 1859, these opposing parties met at the Concord waterfront to board a "steam-tug." This was an unusual vessel for a river where commercial traffic associated with the Middlesex Canal had died a decade earlier. I imagine the air scented with musky odors from the meadow to the north, sunlight sparkling on rippled water, dogs barking, roosters crowing, wagon wheels rolling, cattle lowing, horses whinnying, frogs croaking, and insects humming. The dog days of August had arrived with their blood-orange sunsets. Cow patties desiccated in the dusty sun.[49]

Stepping down from carriages—or perhaps walking over from the railroad station—were five members of a joint special committee appointed by the legislature of the Commonwealth of Massachusetts. They were here to inspect the valley in late summer, when the river stage was at its lowest. There were two state senators, Senator Parker from Norfolk and Senator Bowerman from Berkshire, and three state representatives, Representative Russell of Sunderland, Representative Chase of Salem, and Representative Wrightington of Fall River. Accompanying them was an entourage of farmers, factory owners, town selectmen, attorneys, and perhaps members of the press.

Strong circumstantial evidence suggests that Thoreau was watching them from a safe distance. His journal indicates that he was in town all morning. This would have been the town's most newsworthy event of the summer. This was the culmination of a controversy he'd been thinking about for at least fifteen years and working on for weeks. This date coincides with an abrupt change in his summer research schedule. An earlier journal entry refers to the time when the "Commissioners were here," indicating he was present during their first visit.[50]

After boarding, the delegation steamed thirteen miles up the Sudbury River in a generally southerly direction. Cruising easily beneath the arches of nine bridges, they churned their way through weedy sloughs where hay wagons had once rolled across clean sandbars. At Fairhaven Pond they probably spotted "one-eyed" John Goodwin and gun-toting George Melvin, denizens of the river outback who subsisted off its resources, and whose activities Thoreau closely followed as part of his wild river life. At the upstream limit in Wayland, the committee saw scenery resembling a faux Florida Everglades. Coarse and reedy plants such as cattails and bulrushes emerged from the water where, one century earlier, Minutemen had stood on firm ground to swing their scythes through herbaceous hay.

Turning around, the entourage steamed downstream in a generally northerly direction, floating over bedrock scour holes so deep that only Henry Thoreau had measured them. Deepest was Purple Utricularia Bay, named for a beautiful patch of carnivorous plants that prey on aquatic larvae and zooplankton. That hole had been drilled out by subglacial torrents at a place where the Sudbury River was forced to zig out of its original channel before zagging back to it north of Concord. They also cruised by the bank entrance holes for the North American river otter. Though Henry never saw one, he used field evidence to reconstruct their playful scenes of sliding on snow and their fishy feastings along the edge of river ice. The otter was to Henry's river life what the loon was to Walden Pond: an evasive, fish-eating, deep-diving vestige of a formerly abundant fauna that he could never quite catch.[51]

After cruising past the triple point confluence at Concord, they held a steady downriver course toward Billerica. Within minutes they drifted beneath the site of the Old North Bridge, where a replica built in 1875 (and later re-built) has since become a shrine to American liberty at the epicenter of Min-uteman National Historical Park. Before being replaced, this was the "rude bridge that arched the flood," made famous by Ralph Waldo Emerson's poem "Concord Hymn." Its first stanza is still being recited by schoolchil-dren today. On April 19, 1775, the "embattled farmers" unfurled their flag "to April's breeze" and "fired the shot heard round the world." The second and third stanzas of the poem, however, are virtually unknown. Yet within them are apt descriptions of the valley known as Musketaquid.

> And Time the ruined bridge has swept
> Down the dark stream which seaward creeps.
>
> On this green bank, by this soft stream,
> We set today a votive stone.

"Ruined bridge" was a good call by Emerson, who knew that Concord's eight wooden bridges were frequently swept away. There were twenty such bridges in all the valley. Rising floodwaters from snowmelt freshets, in combination with jagged ice floes and uprooted trees, battered the piers and causeways in an attempt to return the river to its original condition. The river "seaward creeps" with a current so slow that Nathanial Hawthorne claimed to have lived on its banks for three weeks before learning which way it flowed. This river is indeed "dark" for most of the year, stained by the natural tea of swamps, bogs, and marshes, and rendered turbid by suspended clay and colloids. This sluggish organic richness—so different from the bubbling clarity of local brooks—gave rise to what Emerson called a "soft" and verdant "green bank" so fertile that it undergirded the colonial economy.

Steering left at a bedrock bend above Jug Island, the political tugboat chugged northward through the towns of Bedford and Carlisle in a wide, canal-like reach that traces an old geological fault line. Reaching Billerica, the delegation floated over the valley's natural outlet, a bedrock channel called the Fordway. The bottom there was littered with colonial artifacts because this had been an important ford across the river for two centuries, with local residents crossing in water no deeper than the axle of a wagon wheel. Soon the commission members were within the reservoir impounded by the dam below. For half a century its waterline had been marked by an iron bolt hammered into bedrock. After inspecting the dam and its timber flashboards, they turned around and steamed back upriver for hearings in the Concord courthouse, scheduled to begin the next day. Testimony continued nearly nonstop there for morning, afternoon, and evening sessions until October 14. Hearings resumed on October 19 in Boston, and continued until adjournment on December 22. The sum total of transcripts and attached evidence provides the largest and richest source of river history ever compiled in one document.

River Project

In his pathbreaking *The Days of Henry Thoreau* (1962), Walter Harding concluded that Thoreau's lecture "The Succession of Forest Trees," which he gave at the Middlesex Agricultural Society's annual cattle show and which he subsequently published, was his "major contribution to scientific knowledge." Historian Donald Worster concurred in 1977, reporting that the lecture was published not only in the society's *Transactions* for that year but also in the New York *Weekly Tribune*, the *Century*, the *New-England Farmer*, and the *Annual Report of the Massachusetts Board of Agriculture*. Since then, this touchstone of proto-ecology has been highlighted by all of Thoreau's more recent biographers. Oddly, it begins with the whimsical "Every man is entitled to come to Cattle-show, even a transcendentalist," and ends with the blatantly anti-scientific "Surely, men love darkness more than light."[52]

I accept the common claim that "The Succession of Forest Trees" is Thoreau's most important published work of science. But if we consider his unpublished work and use the criteria of analytical rigor, complexity, and scope rather than bulk, his best science is undeniably his river project of

FIGURE 8. Envelope with numbers. From Henry David Thoreau Papers,
1836–[1862]). Envelope addressed to "Mr. Henry D. Thoreaux" and postmarked
Boston June 18, 1859, was used for a numerical table comparing two data sets of
interbridge distances. Courtesy of the Concord Free Public Library.

1859–1860. If we desire to know Thoreau in all his manifestations, not just
those we find most accessible, then we must be willing to slog our way through
this quantitative and technical material, however dense it might be. Unfortunately, Bradford Torrey and Francis Allen, editors of the first published edition of the *Journal*, the 1906 Houghton and Mifflin edition, decided to excise
much of this material, having been "assured on expert authority" that these
data "are now without scientific value."[53]

Their anonymous authority was wrong. For example, it was Thoreau's stage data for June 25–August 15—accurate to the nearest sixteenth of
an inch—that showed river behavior being syncopated with human affairs
to the point where the very fishes follow "man's religion." That authority left
Thoreau's most quantitative work nearly invisible during his canonization as
an American writer. Subsequent scholarship was crippled by that editorial
decision.[54]

"The most serious limitation of the 1906 Edition is its incompleteness,"
concluded Elizabeth Witherill, William Howarth, Robert Sattelmeyer, and

Thomas Blanding, editors of volume 1 of the *Journal*'s Princeton edition. They were referring mainly to the early journal before 1851, from which Thoreau clipped large blocks of text for his literary projects. My main concern with its "incompleteness" is with the editorial culling of technical material from the later journal (1859–1860), when Thoreau was using his journal more as a data repository and place for technical interpretations than as a place to practice good writing. In fact, much of the writing for this interval is downright awful. Tables of numerical data involving the height, velocity, discharge, direction, source, and temperature of water are interspersed with scientific minutiae and ultra-specific interpretations. Consequently, Thoreau's river project was virtually ignored by twentieth century scholars. It received only two paragraphs in Walter Harding's authoritative biography, *The Days of Henry Thoreau*. Lawrence Buell, with a goal of outlining Thoreau's projects, gave it one brief footnote in *The Environmental Imagination*. Robert Richardson's otherwise masterful biography, *Henry David Thoreau: A Life of the Mind*, left it out entirely. Laura Dassow Walls didn't mention it in her remarkable reexamination of Thoreau's science, *Seeing New Worlds*, nor in the journal excerpts she chose for *Material Faith: Henry David Thoreau on Science*.[55]

Twenty-first century scholarship engages the river project more directly, notably Sarah Luria's article "Thoreau's Geopoetics," Daegan Miller's *Witness Tree*, and Patrick Chura's *Thoreau, Land Surveyor*. Though all provide an overview of what Thoreau did, and interpret his engagement in literary terms, none present the work itself, by which I mean the scientific results. For these literary scholars, Thoreau's river project provides an example of something Lawrence Buell recognized: "In the betweenness of mapmaking and place-bonding environmental writing locates itself." But for physical scientists, the river project has been an empty hole.[56]

Happily, the Princeton editorial team and the Thoreau Society have posted online draft transcripts for these excised portions of the relevant manuscript journals: 14, 15, and 16. Unfortunately, large blocks of these transcripts border on incomprehensible to readers unfamiliar with channel hydraulics, fluvial geomorphology, and watershed hydrology. Possessed of this experience, I decoded and interpreted twenty-nine data sets, converted them to tables, graphs, and appendices, and posted them to an online repository. Based on my analysis of these overlapping data sets, I sequence Thoreau's tasks, paraphrase and quote his observations, and explicate his results in three successive chapters.[57]

My hope is that Thoreau's river report will eventually be published as a carefully edited verbatim transcript of text and tables, in a way similar to the gorgeously rendered *Wild Fruits* and *Faith in a Seed*. What makes Thoreau's river project so remarkably different from these botanical studies—and from his still-unpublished work on phenology and Native American ethology—is that it's quantitative and analytical, rather than descriptive and taxonomic. And it's even more geocentric than *Walden*[58]

IN HIS FUNERAL EULOGY, Emerson remarked that Henry's "private experiments" on the river reached the same "result of the recent survey of the Water Commissioners appointed by the State of Massachusetts . . . several years earlier." This is certainly true for Thoreau's covert monitoring of the river stage during the dam pool drawdown experiments dating to August 1859. But stage monitoring was only one small fraction of his much larger project. To present (rather than describe) it for the first time, I combine four of his works—scroll map, statistics of bridges, *Journal,* and archived papers— with two massive published government reports. Largest and most relevant is the Joint Committee's 1860 report, which reprints relevant court documents dating back to earliest colonial times, prints first-person anecdotes about river change dating back to the American Revolution, transcribes weeks of hearings, and reports scientific data. The other government report is a technical engineering study with approximately 35,000 data entries printed in 1862 by a politically appointed commission. Its chief value is a cornucopia of data so good that the subtleties of watershed hydrology during Thoreau's river years can be readily interpreted.[59]

Together, these six primary sources document a watershed ubiquitously altered by Anthropocene changes, and prove that Thoreau understood these changes better than anyone else known to have been involved. Of these impacts, the most widespread were the nearly complete deforestation of the watershed for market agriculture, cordwood, charcoal, and timber, and local extinction of beavers. Most concentrated was the erection of the Billerica dam, which submerged the outlet of Musketaquid: the dam's crest is higher than the bottom of the river for twenty-five miles upstream. Most salient were the sprawl of the transportation infrastructure of roads, canals, and railroads; the building of eighty-two dams for at least fifty factories within the watershed;

the excavations for mines and quarries; and the wanton extinction of much of its wildlife.

The abrupt, early loss of beaver dams in headwater streams enhanced the volatility of streamflows and raised sediment production. The progressive loss of forest made the Concord River more likely to flood during heavy rains, tributary brooks more likely to run dry by midsummer, the lower Assabet to have higher base flows, and Musketaquid to remain wetter in the valley bottom. Thoreau was well aware of this transition. Reservoirs behind mill-dams reallocated the discharges and timings of natural flows and trapped the sediment needed to sustain the meadows, which ultimately enhanced their wetness. Certain reaches became clogged with sediment pollution. The climate was changing; after the 1820s, the trend was toward increasing wetness, with the number of fair summer days decreasing. River ice thinned as the Little Ice Age waned. Dissolved nutrient pollution from manure and eroded topsoil from "run-out" pastures enhanced the growth of aquatic weeds, which choked the river at narrows and backed up the flow. Bridges and causeways segmented the once-continuous Musketaquid into a series of flat-water reaches separated by riffles beneath bridge abutments. The valley was dammed not only at both ends but in the middle as well. There, at the T-junction where the Assabet River met Musketaquid, the combination of higher water backed up by the Billerica dam and sediment pollution from the Assabet raised pre-existing bars to the point that they became natural "earthen" dams, which in turn backflowed the meadows of the upper Sudbury River.[60]

In his compelling book *America as Second Creation,* historian of technology David Nye puts a positive spin on this Anthropocene conversion. "The axe, the mill, the canal, the railroad, and the . . . dam," he writes, "stand at the center of stories about how European-Americans naturalized their claim to various regions of the United States." Indeed, the environmental history of Thoreau country marched in lockstep with the sequence Nye outlines. In 1835, Alexis de Tocqueville summarized the zeitgeist in one sentence: "The American, the daily witness of such wonders, does not see anything astonishing in all this." Importantly, Thoreau's lifespan brackets both sides of Tocqueville's published summary by at least fifteen years. Looking back from near the end of his life, Thoreau wrote: "I have never got over my surprise that I should have been born into the most estimable place in all the world, and in the very nick of time, too." His "nick" was the great acceleration of New England's Anthropocene conversion. The most dramatic changes in river history were launched when he was a barefoot boy dreaming of traveling the world via the scows of the Middlesex Canal. Luckily for us, they peaked when he was a mature observer and an extraordinary communicator. Unluckily, Thoreau died early, leaving his pioneering river science unpublished.[61]

Today, every element of Nye's sequence—axe, mill, canal, railroad, dam—would require some sort of environmental impact assessment. Back then, however, individual landowners, corporations, and government entities acted alone, hoped for the best, and ignored whatever happened beyond their own place, time, and position in the cascade of change. The environment itself had no legal standing, as it does today. Broadly speaking, the cultural context of the flowage controversy was not unlike the global climate change debates during the last quarter of the twentieth century. Even though the impacts were self-evident and alarming to Thoreau's contemporaries, the statistical proofs were subject to challenge, and complex causal relationships had not yet been worked out. Even worse, the legal system depended on precedent, making it difficult to look toward an uncertain future.

The flowage controversy climaxed five decades worth of lawsuits, providing an opportunity for a longitudinal case study of environmental law. To my knowledge, it is the closest nineteenth-century analogue for what, in the twentieth century, became known as environmental impact assessment. Nowhere else did such a low dam impact such a vast area, prompting a statewide analysis with national implications. Though ostensibly a legal case about water levels, the economic, health, and social ramifications of the dam were also considered in great detail. And as with modern environmental assessments, several alternatives to the main proposal were considered.

THOREAU WANTED NO PART IN THE LEGAL PROCEEDINGS. Why? Patrick Chura claims that Henry stayed away because of a vested financial interest: "his recognized need to carry out 'neutral' surveying work." This interpretation suggests that Thoreau, as a surveyor for hire, needed to be seen either as unbiased or on the side of industrial development, where the real money was moving. Historian Brian Donahue reached a conclusion more in sync with my own: that his clients "were both enthusiastic agents and perplexed victims of economic imperatives in the exploitation of land, and apparently never quite understood the connection. Perhaps only Henry Thoreau, passing quietly up the river in his rowboat, had a real grasp of that." This book examines Donahue's illuminating speculation that the valley farmers were unknowingly entangled within a larger environmental transformation that they, themselves, were helping to create. It portrays a legal system unable to resolve a conflict based on grievances that demanded a binary outcome of yes or no: yes, the dam was the main problem, or no, it was not. In contrast, Henry understood that the linkages between appearances and causes were complex, if not unknowable, and that whether the appearances were good or bad was a value judgment. In his mind, the protagonists (with whom he sympathized) and antagonists were less opposing parties than different components of a single, larger natural system.[62]

To imagine Henry on the witness stand under these circumstances is to imagine a tragicomedy. He had been an expert witness several years earlier when summoned to testify, against his will, for someone flooded by a different dam. His side lost. Shortly thereafter he wrote: "I think that the law is really a 'humbug,' and a benefit principally to the lawyers." The lawyers, however, weren't being irresponsible. Rather, water law was simply lagging behind watershed science. And Thoreau was ahead of both curves.[63]

The message for us is to let the science lead the law when it comes to environmental management. Thoreau stepped aside. We cannot.

1 | Moccasin Print

TRAPPING, FISHING, AND HUNTING ALONG THE RIVER were everyday activities for young boys growing up in nineteenth-century Concord. This was certainly the case for young Henry David Thoreau, whose expertise with a "box-trap, fish-hook, or flint-lock shotgun" was admired by his young friend Edward Emerson, son of Ralph Waldo. Thoreau "bragged that he could carry a rifle all day without finding it heavy." By his early thirties, however, he had given up these blood-sport activities for archaeology. This type of hunting gave him durable relics that pleased his imagination. He used them to time-travel from his epoch back to the preceding one.[1]

"I feel no desire to go to California or Pikes Peak," he wrote during the start of his river project. "But I often think at night with inexpressible satisfaction & yearning—of the arrowheadiferous sands of Concord. . . . This is the gold which our sands yield." Thoreau recalled spending "whole afternoons, esp. in the spring,—pacing back & forth over a sandy field— looking for these relics." This gold was created not by the hydraulic mining and sluice boxes of western placer districts but by the more potent human alterations of land deforestation, stump-pulling, harrowing, and finally plowing with ox teams. "Even though the thickest woods have recently stood there," Thoreau remarked, "these little stone chips made by some aboriginal fletcher are revealed."[2]

In entry after entry throughout his journal, Henry described his habit of springtime archaeological reconnaissance not only "in corn and grain and potato and bean fields, but in pastures and woods, by woodchuck's holes and pigeon beds, and, as to-night, in a pasture where a restless cow has pawed the ground." The key to these discoveries was land disturbance. The old-growth forest soil with its dense weave of roots had to be destroyed and the surface washed by spring rains. Inevitably, this produced what Henry called the "perennial crop of Concord fields. If they were sure it would pay—we should see farmers raking the fields for them." In some places, arrowheads were so

FIGURE 9.   Paleo-Indian
point. Small fluted projectile
point (4.3 cm) composed of
red banded rhyolite and
assigned to the late
Paleo-Indian period,
ca. 10,000 BP. Both sides
of one projectile point.
Concord, Dakin's Farm
at Dakin's Brook and the
Assabet River, Concord,
Massachusetts, Paleo-Indian
to the Late Woodland
(AD 1000–1650) periods,
the collection was formed
by Adams Tolman between
ca. 1888 and 1920; Stone. Gift
of Mrs. Adam Tolman, 1990.026.2249.
Courtesy of the Concord Museum.

common they littered the ground as if "it had rained arrowheads. . . . Like the
dragon's teeth which bore a crop of soldiers, these bear crops of philosophers
& facts—& the same seed is just as good to plant again."[3]

Whenever I read such passages, my thoughts turn immediately to the
value of stone artifacts as trace fossils for the geological epoch preceding his
own, the Holocene. Indeed, Concord's postglacial prehistory is marked by a
widespread and virtually indestructible litter of arrowheads, lost tools, and
debitage, the small stone flakes chipped away during tool manufacture. Gradually they accumulated on the black humus of the interglacial soils. Slowly they
worked their way downward through the mulch, coming to rest on mineral
soil. Above them and mixed with them were trace fossils of Henry's epoch,
the Anthropocene: the sawdust and pollen of European plants in wet places
and the "small things forgotten" of everyday Puritan life, the clay tobacco pipes,
broken dishes, occasional coin, and stone scratched by a plow.[4]

When Thoreau paced back and forth in cultivated fields, he was finding
what archaeologists call surface sites. In them, there is no layer-cake context
for sorting out different horizons. A rifled bullet from Thoreau's era, a flint-
lock musket ball from the eighteenth century, and an arrowhead from the late

Archaic Period can all be found in the same handful of dirt. Such artifacts are mixed because they can be let down from above by erosion, worked up from below by heave, or left in place where they were dropped. They cannot be linked to specific layers or horizons where they might be put in context and dated.

Only after thousands of surface artifacts have been recovered, examined for distinguishing features, arranged into a technological series, and linked to datable subsurface sites can archaeologists write a scientific prehistory of human habitation on the landscape. Thoreau's large collection of more than nine hundred Indian artifacts helped contribute to this process, though the stratigraphic story had not yet been worked out. But during the twentieth century, his private collection, dozens of others, and those recovered from professional scholarly excavations have given us a framework—chronological, geographical, technological, and spiritual—for interpreting the human story. Throughout New England, this Holocene story is divided into discrete blocks of time characterized by diagnostic artifacts. The Paleo-Indian Period (12,000–9000 years before present; BP) is known for its fluted points. The pre-ceramic and pre-horticultural Archaic Period is divided into early, middle, late, and terminal (9000–2700 BP). The Woodland Period is similarly divided (2700 BP to circa 1600 CE).[5]

The more recent story of Native Americans is part of the succeeding Anthropocene epoch. Technically, this predates the settlement of Concord because products of European manufacture began showing up in sites long before the Pilgrims landed at Plymouth in 1620. For example, on their first land reconnaissance, before the *Mayflower* crossed the bay to reach Plymouth, Myles Standish and his Pilgrim followers excavated a copper kettle from a burial mound in Truro, on outer Cape Cod. Within the Concord River Valley, Shirley Blancke of the Concord Museum has devoted a lifetime to curating Musketaquid collections and placing them into this regional context.[6]

Significantly, Concord's oldest projectile points are assigned to the Paleo-Indian Period based on their typology. These fluted points are estimated to be roughly 10,000 years old. When they were dropped on Concord soil, ice sheets still lingered over much of Canada, and the local ecosystems were still shifting away from their boreal affinities to pine-oak forests. These earliest entrants to New England are referred to as "Pioneers," likely nomadic hunters passing through the valley on a journey we know almost nothing about. Their artifacts are dated and interpreted by correlation with the Bull

Brook site in Ipswich, Massachusetts, where Paleo-Indians foraged a remark-ably different seashore and hunted caribou further inland.

Was their landscape pristine? No. That term is a fallacy because it pre-sumes some original, timeless condition followed by some disturbing change with a negative connotation. Relentless, value-neutral change has always been the story of the earth's history.

That's why the geological time scale—eon, era, period, epoch, age—is based on fossil turnovers. For the first few millennia after the ice sheets, Pleis-tocene New England was characterized by tundra and boreal communities without a known human presence. Holocene New England, which began about 12,000 years before the present, featured a charismatic fauna that in-cluded humans of uncertain genetic and linguistic affinities. Thoreau called the human part of this fauna "Indians," a misnomer that remains in use today, and the nonhuman part the "nobler animals." He listed them as "cougar, panther, lynx, wolverine, wolf, bear, moose, deer, the beaver, the turkey, etc., etc." He left out the river otter because it was still present, though hovering sight unseen on the edge of local extinction.[7]

Thoreau, of course, could not date his earliest stone artifacts as we do today, with cosmogenic nuclides such as ^{10}Be or ^{26}Al, optical thermolumines-cence, or radiocarbon accelerator mass spectrometry (^{14}C-^{12}N). Rather, he grouped his artifacts by appearance or style of manufacture, which changed greatly from the "Pioneers" of the Paleo-Indian Period to the "Settlers" of the Archaic and early-to-middle Woodland Periods to the "Farmers" (more prop-erly horticulturists) of late Woodland time.

Thoreau found a great variety of artifacts. Most often it was "a new crop of arrowheads. I pick up two perfect ones of quartz, sharp as if just from the hands of the maker." Sometimes he found bodily remains—for example, "a rib and a shoulder-blade and kneepan." Or "soapstone pottery . . . one side of a shallow dish." Or the broken fragments of mud-fired urns, like the one he re-assembled in a journal sketch as if in a museum monograph. Sometimes the prize was a finished piece of art, which confirmed for him that these an-cient peoples were spiritually minded: "a pestle" decorated with "a rude bird's head, a hawk's or eagle's, the beak and eyes . . . serving for a knob or handle . . . a step beyond the common arrowhead and pestle and axe. Some-thing more fanciful, a step beyond pure utility." Or an "elliptically rolled" hammerstone "evidently brought from the sea-shore." Or a slate clam opener that resembled a putty knife decorated with hachured lines.[8]

Paleopsychology is something Henry practiced. Even the tiniest stone flake glistening in the sun yielded "a thought. I come nearer to the maker of it than if I found his bones. . . . No disgusting mummy, but a clean stone, the best symbol or letter that could have been transmitted to me . . . It is no single inscription on a particular rock, but a footprint—rather a mind-print—left everywhere, and altogether illegible. . . . They are not fossil bones, but, as it were, fossil thoughts, forever reminding me of the mind that shaped them." "These little reminders never fail to set me right. . . . They are at peace with rust. . . . [T]he arrowhead shall, perhaps, never cease to wing its way through the ages to eternity." Indeed, it was the durability of these stone tools—flint, chert, hornfels, rhyolite, basalt, slate, soapstone, et cetera—that made them such good stratigraphic markers for the epoch before tools were made of forged metal.[9]

In addition to bones and stones, Henry also found built structures. Beneath Concord's Mill Dam was an ancient fishing weir identified by the birch logs diverting the stream. In the Assabet he found "stones strewn beneath" the bank "in a low wall, as if they had helped form an Indian weir." Thus, the most historically disrupted reach of his three rivers was already being modified in prehistory. In the sluiceway of Pole Brook, he "found an eel-pot or creel, a wattled basket or wicker-work, made of willow osiers with the bark on, very artfully. It was about four feet long." After a lifetime of reconnaissance, Thoreau had mentally mapped out the location of Native American villages and the floor plans of their "wigwams," always near the river, yet always on slightly elevated, consistently dry soils. Though the Native American presence on the physical landscape was ubiquitous, it was a very light touch, orders of magnitude less impactful than that of Euro-American culture.[10]

Nobody else in the valley had a clearer vision of the Holocene epoch than did Thoreau. His archaeological work helped put his world——a bustling agricultural, industrial, and market town with an 1850 census population of 2,249 souls—into its proper perspective.

Musketaquid

Since his boyhood, Henry had puzzled over something very unusual about the Sudbury River. Most of the alluvial valleys he knew from experience (the Charles, Nashua, Merrimack, Hudson, and so on) and from literature (the Mississippi, Euphrates, Indus, Nile, Amazon, et cetera) widen toward their mouths, their floodplains enlarging. That's where meanders have the broadest range, and where reservoirs behind dams are widest and deepest. Weirdly, this pattern was reversed for the Musketaquid, where the widest and rankest meadows lay far to the south in Wayland, about twenty-five miles upstream of the outlet at the Fordway. Wetlands there were so unique that they were given a special name: the Beaver Hole Meadows. It was in these southernmost towns, farthest from the Billerica Dam, that the meadowland farmers complained the most.

Thoreau found a clue to this mysterious reversal when talking with a workman at Ledum Pool, a low swamp linked to lower Nut Meadow Brook, which in turn is linked to the lakelike portion of the Sudbury River. Below the bottom of the swamp there were "old flags," a type of flowering plant indicating pondside conditions. Below that "the mud was twenty feet deep" and showed "three growths of spruce, one above another." And below that was "a hard-pan with iron in it." This sequence, he recognized, was a compelling story of intermittent submergence during Holocene prehistory.[11]

Deep sediment was also exposed elsewhere in the valley, especially in brick pits. At "E. Hosmer's muck hole," in the Great Meadow, he found a basal peat deposit that could be stripped up like "bark into long pieces, three quarters of an inch thick and a foot wide and two long." This sort of "felted" peat—especially when sphagnum moss is present—is diagnostic of submergence. And beneath that peat was the "blue" clay of the underlying glacial lake sediments. During one notable drought, Thoreau saw this same lake sediment widely exposed along the Concord River in Bedford for more than a mile. There it was so rich in glacially pulverized silt and clay, and so poor in organic matter, that it curled up when desiccated, like the giant mud cracks of the alkaline salt lakes of arid areas.[12]

Thoreau did not understand the details and dates of the geological story we know today, but he did get the basics correct. His sequence began with the "diluvial" hardpan at the bottom (during his era, "diluvial" was the vernacular term for the glacial epoch). When the ice sheet receded from Concord about 16,000 years ago, a vast glacial lake lay over the valley, its bottom accumulating the blue clay of Bedford Meadow. With northerly drainage blocked, this lake drained southward through an outlet in Weston. Eventually the ice dam to the north failed, allowing the lake to drain north toward the Merrimack, likely in one or more catastrophic floods. Surface runoff from the watersheds of the Assabet, Sudbury, and Concord Rivers was finally free to pour down into Musketaquid, gather into a broad proto–Concord River, and flow toward the Merrimack. During a transition period, perhaps a millennium long, the combined flow migrated back and forth across the former lakebeds in windswept braided sandy channels, literally paving the way for the meadows to come.[13]

To the indigenous Native Americans, this ancient lakebed, when vegetated and stabilized, became known as Musketaquid, or "grass-ground river," named for the broad natural meadows of succulent grass flanking the channel on both sides. On one cold, clear January day with perfect ice-skating conditions, Thoreau skated twice its full length (upriver and back, and then downriver and back) on a single day. "I was thus enabled . . . to survey its length and breadth within a few hours, connect one part (one shore) with another in my mind, and realize what was going on upon it from end to end,—to know the whole as I ordinary knew a few miles of it only." He concluded that "it is all the way of one character,—a meadow river, or dead stream,— . . . , crossed within these dozen miles each way,—or thirty in all,—by some twenty low wooden bridges. . . . Thus the long, shallow lakes divided into reaches."[14]

Thoreau's insight was spot-on. Twenty earth-filled causeways across the meadows—and the bridges connecting them—had created a chain of shallow lakes resembling irregular beads on a string. For him, it was a boatman's paradise, a watery necklace laid down on the flat floor of what had been a thirty-mile-long glacial lake aligned by the much older tectonic grain.[15]

Some geographic terminology is necessary to avoid further confusion: Musketaquid is the flat alluvial valley. Its southern (upriver) half is the lowermost sixteen lazy miles of the Sudbury River between Framingham and Concord. Above Musketaquid, the Sudbury River drains typical New England highlands crossed by Interstate 90 near Westborough. The northern (downriver) half of Musketaquid is the thirteen lazy miles of the Concord River between Concord and North Billerica. Below that, the Concord River leaves Musketaquid, crossing the natural bedrock constriction of the Fordway before "falling" to Lowell to meet the mighty Merrimack River near the New Hampshire border. The key idea here is that Musketaquid and the Concord River are not synonymous. The former is the flat former lakebed into which the Sudbury flows, and out of which the Concord River flows.

The Assabet River enters Musketaquid from the west after draining highlands parallel to the western arc of Interstate 495. Unlike the other two named rivers, it's a typical New England stream, with a watershed of forested hills and stony brooks that gather into a pebbly main channel with well-defined banks that meanders within a floodplain. Thoreau clearly distinguished the Sudbury portion of Musketaquid from the Assabet: "We are favored in having two rivers, flowing into one, whose banks afford different kinds of scenery, the streams being of different characters. One a dark, muddy, dead stream, full of animal and vegetable life, with broad meadows and black dwarf willows and weeds, the other comparatively pebbly and swift, with more abrupt banks and narrower meadows. To the latter I go to see the ripple, and the varied bottom with its stones and sands and shadows; to the former for the influence of its dark water resting on invisible mud, and for its reflections. It is a factory of soil depositing sediment."[16]

Six years later, Thoreau explained why these two rivers contrasted so greatly: "The remarkable difference between the two branches of our river, kept up down to the very junction, indicates a different geological region for their channels." By "region" he meant watershed, which is the combined surface area drained by the network of rills, ravines, brooks, and tributaries. The outer limit of both watersheds, and of the smaller tributaries of the confluent Concord River, is a continuous, though irregular, line of rock knobs, moss-covered ledges, and forested ridges. Beyond this line, the fate of falling water is to drain away from Thoreau country toward the neighboring streams, including the Nashua, Blackstone, and Charles Rivers. Inside this line, however, everything flows downward and inward to Musketaquid, and then onward to the global ocean. This has been the case for a quarter billion years.[17]

This drainage basin—the Concord River Valley—was the largest, oldest, and most encompassing thing in his life: a "vast amphitheater rising to its rim in the horizon," a "seemingly concave circle of earth, in the midst of which I was born and dwell," a hilly volume of terrain in which the river "fills up the world to its brim." It defined his sense of place. Bisecting this amphitheater is a ridge, capped by a hill called "Nobscot . . . the summit of the island (?) or cape between the Assabet and Musketaquid,—perhaps the best point from which to view the Concord River valley." This ridge was the most prominent of many resistant rock ridges created when the earth's crust was tilted downward to the northwest by tectonic forces and then etched into topographic relief by removal of the weaker adjacent rock.[18]

The result was a series of "five or six such ridges rising partly above the mist" standing above valleys draining southwest to northeast. Thoreau understood that these ridges were aligned with the tectonic grain of the rock in local quarries and ledges, and that this grain also controlled the courses of his three rivers, and therefore his sojourning preferences. "When I am considering which way I will walk," he wrote, "my needle is slow to settle, my compass . . . does not always point due southwest." On that particular day, though, he did go southwest, following the Paleozoic tectonic grain of his amphitheater in an upstream direction away from Concord society.[19]

In *A Week on the Concord and Merrimack Rivers,* Thoreau describes the broad Merrimack River as "a huge volume of matter, ceaselessly rolling through the plains and valleys of the substantial earth." This reveals his insight that his river was much, much more than its flowing water. To geologists, water is merely the vehicle by which rivers create landscapes. Their channels are natural sewers flushing away the aqueous chemistry of soils and the granular remains of rotten rock. They carry particles: the silt suspended within the water and coarser particles of sand and gravel being sheared over the rippled channel bottom, en route to the sea. They carry salts and rusts dissolved from the minerals created in earth's geothermal furnace. They carry carbon extracted from the air by plants in the form of driftwood, dead leaves, and the tannins of tea-colored water. At Walden Pond, Thoreau was mainly interested in the purity and clarity of its isolated, stationary water. On his rivers, he was interested in the relentless flux of anything and everything through time.[20]

In its last mile, the Assabet leaves its northeasterly tectonic trend, hooking southward and then sharply eastward to empty into the alluvial valley. There the bedrock structure forced the hydraulically powerful Assabet to make a T-junction with the hydraulically feeble Musketaquid. Shortly after deglaciation, the entire watershed of the hilly Assabet was unvegetated. Heavy rains, rapid snowmelt, deep freezing, and strong winds combined to wash away staggering quantities of loose sediment. At the T-junction, so much sediment poured down the Assabet into Musketaquid that it created what is effectively an alluvial fan or delta built across the main valley. The easternmost part of this fan-delta pushed Concord's Mill Creek eastward into a resistant sandy moraine, creating the site of future settlement. One of Henry's best informants, Abel Hosmer, described the upper part of this ancient fan-delta as "the whole of that interval covered with sand." Below this were "coarse stones which look like an old bed of the river."[21]

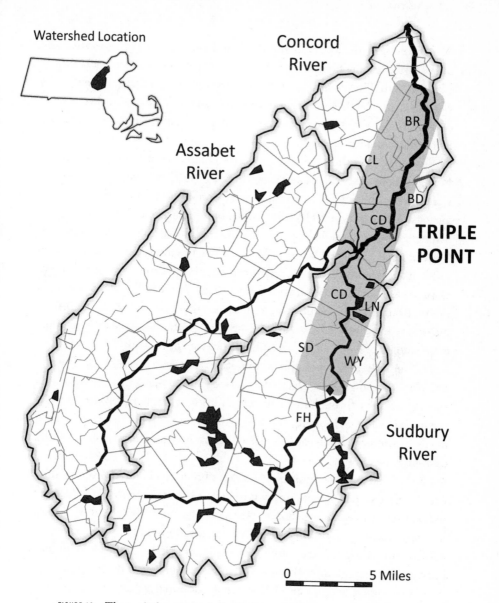

Watershed Location

Concord
River

Assabet
River

BR

CL

BD

CD

**TRIPLE
POINT**

CD

LN

SD

WY

FH

Sudbury
River

0   5 Miles

FIGURE 10. Thoreau's three watersheds. Outlines of the Assabet, Sudbury, and
Concord River watersheds (thin black lines) and their main channels (thick
black lines). Principal tributaries shown by light gray lines. Large water bodies
are black. Town boundaries shown by darker gray lines with the names of
some abbreviated (BR, Billerica; CL, Carlisle; BD, Bedford; CD, Concord;
LN, Lincoln; SD, Sudbury; WY, Wayland; FH, Framingham). The extent
of Thoreau's scroll map is solid gray. North is vertical. Base, U.S. Geological
Survey.

Eventually the loose silt and sand on upland hillsides was washed and blown away, leaving behind a patchy veneer of granules and stones known as a lag horizon. This greatly diminished the supply of mineral sediment to the valley bottom. All three rivers and their tributaries began to run clear, except during floods. Vegetation quickly spread over the landscape like weeds on a fallow field: first tundra, then brush composed of heath and willow, then coniferous forest, and then the rich, moist deciduous forest encountered by the Puritans. Pollen records make it clear that early postglacial forests of Concord were always shifting in composition and dominance, and had no modern analogs. By the middle Holocene, about 6,000 years ago, the time of rapid adjustments to land and forest were over, though the details remained in flux. Organic soils thickened beneath the spreading upland vegetation. Biological nutrients became available, especially carbon, phosphorous, nitrogen, and potassium. Meanwhile, the lakebed silt and clay of the valley floor remained poorly drained. The result was a rich association of wetland plants, mainly mosses, sedges, horsetails, and grasses. For Thoreau, the whole valley became a "broad moccasin print . . . a fertile and juicy place in nature."[22]

The author of that sentiment wondered how such a vast swath of succulent plants could grow on a lawnlike surface without being invaded by woody species. Midway through his river years, in August 1856, he asked: "What is the use, in Nature's economy, of these occasional floods in August? Is it not partly to preserve the meadows open?" Here he uses the phrase "Nature's economy" as we would use the word "ecology" today. Later he would prove to himself that meadow health was indeed a function of both the annual dose of nutrient from predictable spring freshet floods and the timing and duration of less predictable summer floods. At least once every few years there must be warm water standing long enough on the meadows to kill the seedlings of shrubs and trees being dispersed. Yet at the same time, ambient conditions must be dry enough after those inundations to allow free drainage, so that meadow grasses, sedges, and herbs can predominate, rather than rougher aquatic reeds and rushes. What Thoreau lacked was an explanation for the long-term persistence of this unusual hydrology.[23]

That explanation was discovered in 1905 by a Harvard graduate student in geology named Richard P. Goldthwait. He showed that the floor of Musketaquid had been uniformly tilted down to the south by about 4.5 feet per mile, a tilt an order of magnitude greater than the present northerly slope of the river surface. When glacial lakes drain, the axial river usually cuts down through the soft lake-bottom sediment to create a suite of well-drained flanking terraces above the modern floodplain. The whole of the Merrimack follows this pattern. This didn't happen in Musketaquid because the late-glacial downcutting stream encountered a strong granite ridge that became its natural outlet at the Fordway. Simultaneously, the continental crust beneath New England was being tilted southward by rebound of the land to the north, which followed the removal of the enormous weight of the ice sheet. In the south-flowing Merrimack Valley, this increased the river's power. In the north-flowing Musketaquid, the outlet was lifted up relative to the flat lake basin, comparable to lifting the drain of a bathtub above its bottom. The result was an ancient lakebed covered by a film of water that was deepest toward the south. The submergence sequence Thoreau recognized at Ledum Pool is consistent with what actually happened during Holocene times. With little mineral sediment coming in, and with submergence slow enough to allow plant growth to keep pace, the result was a vast organic bottomland covering an estimated "ten to fifteen thousand acres."[24]

For millennia, the wetland ecology remained in an approximate state of balance. Slight changes in the climate, mineral input, and changes in forest cover nudged the system toward drier or wetter conditions. But it was never dry enough to allow forest to invade, nor wet enough to allow a true reedy marsh to develop. When combined with local reaches of open water, and with drier soils at higher elevation, these meadows created a mosaic of habitats rich in fish, waterfowl, turtles, clams, and mammals that nourished a succession of human groups throughout the Holocene, accounting for the richness of its archaeological record. This environment easily survived the first century of European colonial settlement, when Puritans scythed natural hay from the wetland's surfaces, hauled it into their barns each summer, and used it to feed their cattle; they kept the river free of dams during this time so that the fish they depended on might migrate freely into their nets. But the moist fertile meadows they depended on would not survive the more intense industrial development of Thoreau's century.

THE "GOOD" ARCHAEOLOGICAL SITES Thoreau excavated all came from Musketaquid. All had discrete layers buried and separated from one another. Thoreau describes an example: "The soil of that rocky spot on Simon Brown's land is quite ash colored—(now that the sod is turned up) by Indian fires—with numerous pieces of [char]coal in it." Best of all was the site known as Clamshell Bank, located just upstream from his "boat place." Before being buried by expansion of Emerson Hospital, it was the most productive archaeological site on the river, the "only known large inland shell midden in New England," the site of a summer village dating to the middle Holocene, about 5,000 years before the present.[25]

During Thoreau's lifetime, it was a veritable pharaoh's tomb, laid open each spring by river and gully erosion for him to examine. He visited the site over a span of forty-three years, beginning with an entry on October 29, 1837, which was only one week after starting his journal at Emerson's suggestion. His final entry about archaeology dates to August 22, 1860, only a few months before his fatal illness. Thoreau's first clear description of the Clamshell Bank site and the source of its name occurs in his first publication about nature, "A Natural History of Massachusetts," which appeared in the *Dial* in 1842. "That common muscle [*sic*], the *Unio complanatus*, or more properly *fluviatilis* . . . appears to have been an important article of food with the Indians," he wrote. "In one place, where they are said to have feasted, they are found in large quantities, at an elevation of thirty feet above the river, filling the soil to a depth of a foot, and mingled with ashes and Indian remains."[26]

Clamshell Bank, also known as Concord Shell Heap, was the archaeological bull's-eye of the valley, before being thoughtlessly destroyed. There, a sharp bend in the river sends rising floodwaters against a warm, dry, south-facing hill composed mainly of well-drained sand. The result is a warm amphitheater. Adjacent is a perennial spring of fresh water, created where permeable sand of a glacial lake delta overlies the impermeable lakebed clay. A flat terrace cut into this bank was large enough for a native village, and high enough to avoid flooding. Owing to persistent gully erosion, there was a copious supply of fine sand. When graded by the river current, it provided habitat for the highest concentration of clams on the whole river. One day Henry counted an estimated "16,335 clams to twenty rods of shore (on one side of the river), and I suspect that there are many more." They were so thick that during a severe drought Henry "spent half an hour overhauling" his boat over "the heaps of clamshells under the rocks there." Following a lifetime of experience, he had learned that "clams are chiefly found at shallow and slightly muddy places—where there is a gradually shelving shore. Are not found on a very hard bottom—nor in deep mud." Clamshell Bank was one of few places in the whole alluvial valley meeting these criteria.[27]

Dining on those abundant clams was a locally high population of musk-rats. And dining on both was the "vanished race" of human beings who lived there during the Archaic Period. From his journeys to Maine, Henry had learned that living "Penobscots . . . make a very extensive use of the muskrat for food." Henry concluded that "it would seem that they used the fresh-water clam extensively also, these two peculiarly indigenous animals." The meat of turtles was also important, especially the easy-to-catch snappers, which can be thought of as the alligators of the Musketaquid. During one January thaw Henry learned that "one man counted eighty or more dead, some of which would weigh eighty to a hundred pounds. . . . No wonder the Indians made much of them. Such great shells must have made convenient household utensils for them."[28]

By 5,000 years ago, Musketaquid was as archaeologically rich as it ever would be, perhaps because the coastal wetlands were not yet able to compete in richness, and the interior valleys were reaching their peak postglacial warmth. From the Clamshell Bank site came projectile points, knives, scrapers, mortars, pestles, bowls, pots, awls, and fishing sinkers called plummets, as well as the remains of aquatic fauna, large freshwater mussels, fish, many kinds of turtles, waterfowl, and small mammals. The residues from living and feasting created a layer of limy earth more than a foot thick, a midden of broken shells that Thoreau could see from his boat. Another stratified site lying further up-river at Pantry Brook is known as the Davis Farm site. It too has Archaic occupation horizons up to 7,000 years old.[29]

February 28, 1855, was a special day for Thoreau's archaeological career. That's when he described in great detail the formation of a new ravine opening up at Clamshell Bank, one that would keep on eroding to reward him with treasure for the rest of his life. "I observed how a new ravine is formed in a sand-hill . . . a ravine some ten feet wide and much longer, which now may go on increasing from year to year without limit, and thus the sand is ravished away. I was there just after it began." On August 22, 1860, that ravine rewarded him with the best find of his life, one he had a premonition of finding. "Heavy rains have washed away the bank here considerably—& it looks and smells more mouldy with human relics than ever . . . one foot beneath the surface—& just over a layer some three inches thick of pure shells & ashes,—a gray-white line on the face of the cliff.—I find several pieces of Ind. pottery with a rude ornament on it, not much more red than the earth itself . . . I find in all 31 pieces. Averaging an inch in diameter & about a third of an inch thick."[30]

He then proceeded to piece the pot together and to write a detailed description of its size, method of construction, and pattern of ornamentation, drawing a museum-quality sketch in his journal. It seemed that every trip to Clamshell Bank (which he also called Clamshell Hill, or Clamshell for short) yielded something new: "five arrowheads" on May 25, 1856 and "three arrowheads and a small Indian chisel" on July 5, 1857. The material past was the gift that kept on giving.

The pottery he found was diagnostic of the Woodland Period, which began 2,000 years ago as a successor to the Archaic Period. The very nucleus of European colonial settlement, the Mill Dam, was built on top of a Woodland Period fish weir. In fact, human skeletons were accidentally exhumed during construction in the village center. These were remains of a horticultural people who burned off the drier terraces above the marshes to plant crops of maize, beans, and squash. Based on their material culture, they were likely the antecedents of a cultural group known as the Pawtuckets, who would later deed their lands to the English Puritans. According to historian Brian Donahue, they "wintered in several villages of a few hundred each, located at intervals along the rivers . . . a group of longhouses, pole frames covered with bark, with an extended family group of forty or fifty living in each." The epicenter of their world, the T-junction marked by Egg Rock, coincided with Thoreau's.[31]

By the time of European contact, disease epidemics and the early fur trade had reduced native culture to what Donahue described as "catastrophic disarray." The prevailing view from Thoreau's era saw this devastation as a blessing. Lemuel Shattuck, Concord's first historian, wrote in the early nineteenth century of a great pestilence that had struck the area in 1612. Within a generation the native population had been reduced to five or ten families living below Nashawtuc Hill under the leadership of two sachems, a man named Tahattawan and a woman whose name has not survived in the historic record. "This great mortality," Shattuck wrote, "was viewed by the first Pilgrims, as the accomplishment of one of the purposes of Divine Providence, by making room for the settlement of civilized man, and by preparing a peaceful asylum for the persecuted Christians of the old world." Within one generation of that plague, the natives who survived would sell their land for a pittance in order to survive another year.[32]

Twenty years after Shattuck's racist reflection, Henry Thoreau sadly witnessed the final disintegration of Native American river culture. "I have not seen one [Indian] on the Musketaquid for many a year," he remarked in 1850, "and some who came up in their canoes and camped on its banks a dozen years ago had to ask me where it came from." His final Concord memory of them was particularly poignant: "A lone Indian woman without children, accompanied by her dog, wearing the shroud of her race, performing the last offices for her departed race. Not yet absorbed into the elements again; a daughter of the soil; one of the nobility of the land."[33]

2 | Colonial Village

CONCORD WAS CONCEIVED IN THE EARLY 1630S as an English Puritan plantation to be established in the interior wilderness. That's when William Wood, an adventurous promoter of early settlement, claims to have visited the place. In *New England's Prospects,* his exaggerated promotional tract, he described the future town's main attraction. There were "divers places neare the plantations great broad Medowes, wherein grow neither shrub nor Tre, lying low, in which Plaines growes as much grasse, as may be throwne out with a Sithe, thicke and long, as high as a mans middle; some as high as the shoulders, so that a good mower may cut three loads in a day."[1]

Credit for discovering these rich meadows is also given to Captain Simon Willard, who was brought there by one of his Native American guides during his work with the fur trade in the Massachusetts Bay Colony. Though recent scholarship supports this view, Concord's early historian, Lemuel Shattuck, in 1835 cited "traditional authority" for Wood's claim of first "discovery." Certainly Wood was the first to use the name "Concord" as a marketing ploy for establishing America's first inland river town.[2]

The settlement of Concord (1635) came twenty-eight years after the first attempt at English colonization at Sagadahoc, Maine (1607), fifteen after the Pilgrim settlement at Plymouth (1620), and six after the founding of the Massachusetts Bay Colony (1629). Prior to Concord, nothing could induce settlers to move inland beyond the smell of the salt marsh and the lifeline of the Atlantic shore. Henry Thoreau described the inducement in his famously long opening line of *A Week on the Concord and Merrimack Rivers:* "The Musketaquid, or Grass-ground River, though probably as old as the Nile or Euphrates, did not begin to have a place in civilized history, until the fame of its grassy meadows and its fish attracted settlers out of England in 1635, when it received the other but kindred name of CONCORD from the first plantation on its banks, which appears to have been commenced in a spirit of peace and harmony."[3]

FIGURE 11.   Lawnlike meadow. Historic photograph. Looking toward Egg Rock.
Shows relationship between river, alluvial meadows, and low uplands. No
attribution or date. Courtesy of the Concord Free Public Library.

This sentence foreshadows the bitter discord over how the river should be managed when competing uses began to conflict with one another. Should the meadows be managed for haymaking, submerged for mill power, or left alone to enhance fish habitat? Or should they be ditched for crops, dredged for canals, or mined for muck? "It will be Grass-ground River as long as grass grows and water runs here," wrote Thoreau. "It will be Concord River only while men lead peaceable lives on its banks." That concord lasted just three years, until 1639. Discord continued for another 223 years, until the spring of 1862. Within two weeks of Thoreau's death, and circa the outbreak of the Civil War, his generation of meadowland farmers suffered a final crushing defeat from which they never recovered. By then, and in Thoreau's words, the river was "much abused."[4]

The Musketaquid meadows were an unusual and highly productive type of freshwater marsh. To farmers, it was a "smiling mead" where livestock fodder was free for the scything and the soil firm enough to support oxcarts. Lacking woody vegetation, it wasn't a swamp. Lacking acid-loving heath plants and moss, it wasn't a bog. In places it was rich enough to be called a fen (as in Fenway Park), a label that Henry used often. Except when flooded, it was a place of great economic reward.[5]

Thoreau underestimated the antiquity of Musketaquid's grassy floodplain. Those of the Nile and Euphrates didn't begin to form until about eight thousand years ago, when the last of the large ice sheets disappeared. That's when the rate of sea level rise slowed enough to allow river sedimentation to keep pace. By that time, however, the Musketaquid had been supporting marshes for at least five millennia because the outlet of its alluvial basin was a local bedrock spillway independent of the sea.

Concord was legalized in Boston on September 2, 1635, when a professional soldier named Captain Willard, a Puritan minister named Reverend Peter Bulkeley, and their followers were granted permission by the General Court of the Massachusetts Bay Colony to establish a new plantation at Musketaquid. Their proposed settlement was peopled in Cambridge on July 5, 1636, when the future proprietors gathered to establish a congregation. It was domiciled on September 8, 1636, after the settlers had survived the difficult trek through the forest. Already the General Court was referring to the "inhabitants of Concord," specifying their settlement as the place where three rivers met: the North, South, and Great Rivers. By 1637 the settlers had purchased the land from the few remaining Native Americans. Though the deed has been lost, documents from the General Court indicate that it was freely signed by Tahattawan, a woman referred to as Squaw Sachem, Wibbacowett, Natanquatick, and Carte. Fifty years later, a witness named Jehojakim claimed to have been present at the signing, remembering the price as "a parcel of wampumpeage, hatchets, hoes, knives, cotton cloth, and shirts.'" Apparently, legal permission to settle the land, the official establishment of the community, and the physical occupation of the place all predated its purchase from the rightful owners.[6]

Thoreau poked fun at William Wood by quoting his marketing pitch for the quality of Concord's water: "It is thought there can be no better water in the world, yet dare I not prefer it before good beer, as some have done, but any man will choose it before bad beer, whey, or buttermilk. Those that drink it be as healthful, fresh, and lusty, as they that drink beer." This was pure hyperbole. By Thoreau's time there was talk of tapping Walden Pond to quench the thirst of this bustling riverside town.[7]

In one of his earliest journal entries, Thoreau made it clear that Concord River was the "jugular vein of Musketaquid." Indeed, all rivers are veins (rather than arteries) because they drain fluids back to the heart of the sea. "Nature made a highway from southwest to northeast through this town (not to say county), broad and beautiful, . . . ten rods wide and bordered by the most fertile soil in the town, a tract most abounding in vegetable and in animal life." For the town's coat of arms, he suggested a "field verdant with Concord [river] circling nine times round." This scenario seemed true during the March freshet of 1859. As seen from a nearby hill, "the town and the land it is built on seem to rise but little above the flood . . . as if the distant town were an island. I realize how water predominates on the surface of the globe." At that point he compared Concord to Venice, Italy, which was also founded adjacent to a marsh.[8]

The meadows were nourished by the muck beneath them. Muck is the technical name for an alluvial soil that, in this case, was physically sedimented, chemically fortified, and well watered by annual overflows. Grass rooted in muck was free for the scything, self-fertilizing, and abundant enough to sustain the town for ten generations. These alluvial meadows provided about three times as much hay per acre as did the upland ones, and without the need for watering and manuring. From day one of colonial settlement, these floodplain soils literally grounded the community economy. Other agricultural soils included the drier and less fertile sandy and loamy soils of the upland terraces and hills, which were tilled for crops, planted as orchards, fenced for pastures, and gardened for vegetables. "Fresh meadow hay" was even more important for Concord's colonial husbandmen than it was for their English counterparts because New England winters were longer, meaning cattle were confined in barns longer. Prior to the American Revolution, it provided at least 70 percent and as much as 85 percent of their cattle fodder.[9]

"Give us this day our daily bread." Those words from the Christian Lord's Prayer would have been heard night after night in candlelit colonial homes. Bread made of cereal grains, mainly corn and rye, was the final link in a food chain that is eight links long, beginning with the river. The river gave them muck. The muck gave them hay. The hay gave them cattle fodder. The cattle gave them manure. The manure was used as fertilizer. With fertilizer, the otherwise lean upland soils produced the grain on which their subsistence depended. And finally, that grain gave them their daily bread. The community's debt to the river was confirmed by its spatial layout: "Nearly two hundred mowing lots flanked the watercourses throughout the town." By 1650, "something very much resembling a classical English common field village, in all its intricate complexity, had been set down in the midst of the American forest." Concord's early colonial lifeway was a sustainable, long-term cultural practice derived from those of Old World aristocrats concerned with maintaining resources, particularly wood.[10]

CONCORD'S EARLY DEPENDENCE on the river's bounty came with one significant hitch: sometimes the "great and peevish river" would rise unexpectedly to flood the meadows, ruin the hay, destroy the bridges, and soften the ground so much that livestock sank up to their bellies. On the other hand, the constant threat of imminent flooding helped unify the early settlers by giving them a common enemy to guard against. So when the grass was ready, the entire village worked overtime together, night and day, until the hay was safely secured in barns.[11]

Two centuries later Thoreau described a scene changed little since the first Puritans arrived: "The Great Meadows present a very busy scene now. There are at least thirty men in sight getting the hay, revealed by their white shirts in the distance, the farthest mere specks, and here and there great loads of hay, almost concealing the two dor-bugs that draw them—and horse racks pacing regularly back and forth. . . . Here are many owners side by side, each taking his slice of the great meadow. . . . The completion of haying might be celebrated by a farmer's festival."[12]

One of these surprise floods interfered with the hay harvest in 1636, during the very first year of settlement. After this inauspicious beginning, the proprietors proposed a solution. On September 8 they informed the General Court that they would "abate the Falls in the river upon which their town standeth." The "Falls" referred to the rapids at the outlet of Musketaquid, called the Fordway. Their request came with the stipulation that future upstream settlements help defray the cost for any drainage work done, contributing in proportion to their "charge and advantage." The first subsequent upstream plantation was Sudbery (now called Sudbury, from which East Sudbury or Wayland was also carved), created by the "General Court holden at Boston the 4th day of 1639." Within three years the residents of Sudbury would also be asking the English colonial governors for permission to work with Concord in dealing with pesky floods. On July 15, 1642, they appealed to "his Excellency, William Shirley, Esqr., Capt. Generall and Governor in Chief in and over His Majesty's Province of the Massachusetts Bay in new England," requesting permission to create a board of "Best Commissioners of Sewers." In their day, the word "sewer" referred to any utility used for land drainage.[13]

Though the villages of Concord and Sudbury worked together to abate the "falls," they never succeeded. Despite some digging and blasting, cutting through the bedrock obstruction of the Fordway lay beyond their engineering capabilities. As an alternative, they explored the possibility of diverting floods eastward into the nearby Shawsheen River on the far side of Bedford Flats, believing they could be "turned another way with an hundred pound charge" of black powder. These early accounts show that the early Puritans were as aggressive about physically changing their world as people from later centuries, but lacked the technology to do so. The dream of deepening the channel at the Fordway persisted for an additional two centuries. In August 1859, during the peak of Thoreau's river project, an author identified only as "R" suggested that "the channel be deepened three feet, for one or two miles, at and above the falls, a work which is entirely feasible and practicable."[14]

By September 7, 1643, the mood in Concord had become seriously discordant. "Finding the lands about the town very barren, and the meadows very wet and unuseful," residents sought permission to leave. Permission was denied. They left anyway. Families began to depopulate Concord in

1644: some moving back to England, some to nearby settlements, and many to Connecticut. Those who remained begged to be released from the town charter and be allowed to leave. Humbly petitioning the governor, they complained of the "povertie and meanness of the place" and noted that the land was not "answering the labour bestowed on it," largely owing to the "badness and weetnes of the meadowes," causing "many houses in the Towne [to] stand voyde of Inhabitants." By 1645 the General Court "ordered that no man now inhabiting & settled in any of the s'd Townes (whether married or single) shall remove to any other Towne without the allowance of the majistrates or the select men of the towns." Residents sought official permission to leave through 1672.[15]

Flooding also created serious transportation problems. The first bridge across Musketaquid, built in 1665, was washed away in a storm; another was built the following year on the same spot, now called South Bridge. In 1835 the town's first historian reported: "Six or seven new bridges have since been built on the same spot. . . . They have been often swept away by the floods; and large sums of money were annually raised to keep them in repair." Curiously, these bridges over the north-flowing Sudbury River were being washed out by south-flowing back floods of the Assabet. More generally, bridge damage was largely self-inflicted, because the flood currents were strengthened by more than an order of magnitude owing to the constrictions of causeways and bridge abutments.[16]

Deserted Country

One of the tasks that Henry agreed to do for his clients in 1859 was to interview long-term residents about their historical river memories. After this short-lived oral history project, he realized that valley history could be divided into two basic stages. "200 years ago is about as great an antiquity as we can comprehend or often have to deal with. It is nearly as good as 2000 to our imaginations," he wrote. "It carries us back to the days of aborigines & the Pilgrims . . . beyond the limits of oral testimony, to history which begins already to be enameled with a gloss of fable—and we do not quite believe what we read."[17]

Specifically, two centuries brought him back to 1659. By that time, village life had already been more than a generation old. History beyond that threshold felt more like prehistory, when Concord was "a vast & howling place . . . where a man might roam naked of house & most other defence—exposed to wild beasts & wilder men." This was Thoreau's paraphrasing of Pilgrim governor William Bradford's vision of this interval. "Celebrate not the Garden of Eden," Henry reflected, "but your own." Here he imagines Concord being founded as a colonial Eden, a village nourished by luxurious green meadows—and not to be confused with a "city on a hill," John Winthrop's reference to the Massachusetts Bay Colony as a fortified New Jerusalem.[18]

Psychologically, ancient colonial Concord was more accessible to Thoreau through archaeology than through the old parchment manuscripts of the town hall. So he treated the artifacts of early English settlement as he did those of the Native American Woodland period. "If I did not find arrowheads I might, perchance, begin to pick up crockery and fragments of pipes—the relics of a more recent man. . . . Or I might collect the various bones which I come across." Together "they would make a museum . . . & what a text they might furnish for me for a course of lectures on human life or the like!" Bullets and arrowheads were often found together, "as if, by some unexplained sympathy and attraction, the Indian's and the white man's arrowheads sought the same grave at last." That sympathy, of course, involved the hunting of wild game, the defense of domiciles, and fording rivers at similar spots. These universal human motives of village life transcended race and ethnicity.[19]

Early in Thoreau's archaeological career he took delight in the fact that his world was not built on the ashes of a former one. With more experience he changed his mind. The ruins of the ancient Puritans, he discovered, were nearly as widespread as those of the Native Americans. He could truthfully speak of a European "race now extinct, whose seines lie rotting in the garrets of their children, who openly professed the trade of fishermen, and even fed their townsmen creditably." Having emigrated from English coastal villages, the early settlers remained fishermen well into the eighteenth century, when that lifeway was denied by the first dam in Billerica, which blocked the passage of migrating fish such as shad, salmon, and eels in 1710. Another clue Thoreau found to the earlier, simpler way of life was a bag of pebbles dating to 1626 that he found in a neighbor's house. They were weights for measuring commodities for barter, "all rather dark and ancient to look at . . . I love to see anything that implies a simpler mode of life and greater nearness to the earth."[20]

Thoreau wrote poignantly of the aboveground ruins he found nearly everywhere, including those of his own century: "Farmhouses nearly half a mile apart, few and solitary . . . the still, stagnant, heart-eating, life-everlasting, and gone-to-seed country . . . the apple trees are decayed, and the cellar-holes are more numerous than the houses, and the rails are covered with lichens . . . What must be the condition of the old world! The sphagnum must by this time have concealed it from the eye. . . . Perchance when the virgin soil is exhausted, a reaction takes place, and men concentrate in villages again."[21]

"We walk in a deserted country," he wrote of Musketaquid towns. "This is ancient Billerica (Villarica?), now in its dotage . . . gone to decay, farms all run out, meeting-house grown gray and racked with age? If you would know of its early youth, ask those old gray rocks in the pasture." One year later he "called at the Conantum House" in Concord, then in the process of being converted to soil, and noted, "It grieves me to see these interesting relics, this and the house at the Baker Farm, going to complete ruin." In one fascinating story he tells of the foundation of an ancient house with wooden stands called "horses" in the cellar to support eight liquor barrels. "The first settlers made preparations to drink a good deal—& they did not disappoint themselves." All this was then a complete ruin in the woods.[22]

Abandoned mines also dotted the landscape: "The Cooper mines—the old silver mine now deserted. . . . The bog-iron mines—the old lime-kiln—the place where the cinnamon stone was found." The charcoal mounds for those ancient kilns lay in ruins as well. "Saw to-day . . . a singular round mound in a valley, made perhaps sixty or seventy years ago. Cyrus Stow thought it was a pigeon-bed, but I soon discovered the coal [charcoal] and that it was an old coal-pit," he wrote. "We find no heroes' cairns except those of heroic colliers, who once sweated here begrimed and dingy, who lodged here, tending their fires." Old quarries littered the landscape, especially in Carlisle and southwest Concord.[23]

Most evocatively, Thoreau found sawmills and gristmills so old they had already been abandoned and their foundations overgrown. Once, with Nathaniel Hawthorne, he "reached the ruins of a mill where now the ivy grew and the trout glanced through the raceway and the flume." Later he came to understand that this ruination was a self-inflicted consequence of land conversion. "It is interesting to see near the sources, even of small streams or brooks, which now flow through an open country, perhaps shrunken in their volume, the traces of ancient mills, which have devoured the primitive forest, the earthen dams and old sluiceways, and ditches and banks for obtaining a supply of water. These relics of a more primitive period are still frequent in our midst. Such, too, probably, has been the history of the most thickly settled and cleared countries of Europe. The saw-miller is neighbor and successor to the Indian." Indeed, with respect to their respective trace fossils, the sawdust of the Anthropocene was successor to the debitage of Holocene hunters.[24]

Most ubiquitous, Thoreau noticed, were the old stone walls on vacant farms that proprietors had walked away from. "The oldest monuments of the white settlers hereabouts are probably some dilapidated & now undistinguished stone walls—laid long before Philip's war [1675–1676]—not houses certainly perhaps not cellars—but old unhonored stone walls & ditches—But it was difficult to find one well authenticated. I respect a stone wall therefore." This was also true for the stone walls of his era, the vast majority of which are now aboveground artifacts without written or legal documentation. With great affection Thoreau looked back two centuries to imagine an ancient heroic age of wall builders right in the midst of Concord's second-growth woodlands, being cut once again for fuel to feed the ravenous railroad: "When I see a stone which it must have taken many yoke of oxen to move lying in a bank wall which was built 200 years ago—I am curiously surprised because it suggests an energy & force of which we have no memorials. Where are the traces of the corresponding moral and intellectual energy?"[25]

"The old ways are already gone," he wrote in 1854, having seen these "relics of a more primitive period" of Anglo-American history. Along with sawdust and lost domestic items, these already tumbling walls index the base of Concord's Anthropocene epoch just as surely as the earliest stone projectile points index the base of the Native American Holocene.[26]

Osgood's Gristmill

During early colonial settlement, hinterland towns were isolated communities linked to one another by the rudest roads and oxcart paths. Aside from necessary common buildings such as the meetinghouse, tavern, and general store, every village needed a sawmill to cut the timber for houses and barns, and a gristmill to grind the grain into flour. Mills demanded hydropower, which necessitated the presence of mill seats. These were places on streams where the foundation was strong enough to support a dam and the channel was resistant to erosion. There also had to be a sufficient quantity of water, the space to back it up in, and a fall high enough to power mill machinery. From the very beginning of New England settlements, milldams were indispensable requirements, and thus strongly supported by colonial laws.[27]

Though it was the meadow grass that drew settlers to Concord, it was the mill seat on Mill Brook that nucleated the village. Concord's milldam was seated where the sluggish brook narrowed and dropped down to the level of the river. At this place was an old fishing weir that the natives had used to trap migrating schools. Though it was a poor mill seat for large-scale manufacturing, it was nonetheless developed because it lay immediately adjacent to what mattered most: the Great Meadow, for its hay-producing muck, and the adjacent Great Field, for its grain-producing loam.

Things were very different in North Billerica. There, the discharge from about 350 square miles was gathered into what the early settlers called the Great River, which flowed through a bedrock notch immediately above a rapid known as the Falls. On October 4, 1704, the relatively new Town of Billerica granted a mill seat there to "Christopher Osgood, Junnr., of Andouer, all that neck of land on the West of Concord Riuer, lying between said Riuer and the pathway leading to broad meadow, with the stream at the falls." In addition, the grant reserved a length along the river of "ten pole from the fordway doun said Riuer" for a mill pond "from the foot of the Hill going doun into broad meadow."[28]

The grant was dependent on three conditions: first, that "said Christopher Osgood do, within two years next ensuing the date hereof, Erect and maintain a good grist mill upon said Riuer, at the falls"; second, that the same "said Osgood doth Engage to secure and defend the Town of Billerica from any trouble and charge that may arise for damage that may be don to the meadows of the Towns aboue us by said mill-dam"; and third, "the said land is given & granted to the said Christopher Osgood and his heirs so long as he and they shall maintain a good grist mill at said place, and when said mill ceases, the said land shall Return to the said town of Billerica."[29]

This 1704 document is a dark foreshadow of the flowage controversy. Even before the first village gristmill was built, there were serious concerns that its dam would back-flood the waters far enough to invite challenges from upstream towns. Within six years this actually happened. In 1710 the upstream town of Concord "remonstrated" to the General Court of the Commonwealth against the dam's construction. In response "it was voted, that the Town of Billerica will defend Mr. Christopher Osgood from bearing any charge of the damage in flowing [flooding]" the river and meadows upstream. Osgood built his dam in the midst of this intervillage conflict. We don't know what materials he used because the dam was destroyed, covered, and recycled into the larger, later dams. It was, however, rumored to be leaky and inefficient.[30]

Objections to its presence were immediate. Consider this document from the colonial General Court dated February 22, 1714: "Having received some lines from some of ye Inhabitants of Concord, requesting us that we would choose a Committee to view the River and find out the stopage of the water and consider what may be best done to remove the encumberance thereof." Three days later, "at said meeting," it was "ordered that five people . . . join with Concord to act in said affair." This sounds eerily similar to the makeup of the five-member Joint Committee appointed nearly a century and a half later, in 1859.[31]

The main argument for removing the dam in the early eighteenth century was that it blocked the migration of fish, which were then a critical part of the economy. Concord's first historian, Lemuel Shattuck, presented a before-and-after scenario: "The fish formerly most abundant in Concord were salmon, shad, alewives, pike or pickerel, dace, and some others. . . . The principal fish, which now inhabit these waters, are pike, perch, lamprey and common eel, pout, and several other smaller fish."[32]

Thus it was that Concord River joined a long list of New England streams that became battlegrounds between the value of migratory fish and the value of falling water to power mills. This story is well told by David Montgomery in *King of Fish,* which features the Atlantic salmon, and by John McPhee's *Founding Fish,* which features the American shad. Both of these oily, energy-rich fish were vital to colonial subsistence. Notes Montgomery, "Some authorities describe the pre-contact salmon runs of the eastern United States as comparable to the legendary runs of the Pacific Northwest." Captain John Smith, Pilgrim Thomas Morton, and Puritan promoter William Wood all confirmed the abundance of salmon. Wood claimed that salmon were so abundant they were too "cheape" to sell.[33]

In 1709 the colonial legislature passed a law forbidding construction of new milldams that would block the passage of migrating fish such as salmon and shad. However, it exempted those dams already in place or approved. Thus the Billerica gristmill was grandfathered in because its grant predated the law by five years, even though construction came later. Most of these laws, however, would be ignored and eventually reversed for the sake of industrial development. Hereafter, the date of 1710 is used for the raising of the first dam across the Concord River in North Billerica.[34]

In 1720 the colonial governor and council sent a three-man commission to Billerica to visit the dam and sound the nearby channel. They concluded not only that it blocked fish but also that it "greatly hinders the water's discharging itself." In response, a seven-member Commission of Sewers demolished the dam in 1722 using "force and arms." Osgood sued for trespass but lost the case, and then appealed. Officials allowed him to rebuild on the condition that the milldam be kept open two months per year to let the fish migrate. He didn't comply. Opponents of the dam then sought, and were awarded, a "warrant for a town meeting drawen April 30, 1723," asking "that the stopage and obstruction upon Sudbury and concord river may be removed, which we are informed is upon said river in Bilerica bound, which is a hinderance to the passage of the Fish." Thoreau, who enjoyed perusing the town's early records, may have read of this early uprising.[35]

BY JULY 15, 1742, the colonists had become less concerned about their fishing privileges than about the quality and quantity of meadow hay. That's when they began to report that weedy bars and other obstructions were causing wet meadows further upstream. In that year Joshua Haynes and others petitioned the General Court for authority to help clear the river of weeds because "the many Bars and stoppages which are in the said River" are enhancing "the Flods which hath and Do very often overflow and stand a long time upon our said Meadows." This had nothing to do with the dam at Billerica, which was too low at this stage to make any difference. Thus it seems that early land conversions in the upper watersheds were already beginning to impact the flood regime.[36]

Six years later, in 1748, the executor for Osgood's estate sold his gristmill privilege to Nicholas Sprake, who sold it to William Kidder, who in 1759 sold two-thirds of the privilege to John Carleton. Sometime after the American Revolution, Carleton sold the already clouded legal title for his mill privilege to Thomas Richardson, who built the first large factory on the site. For more than a century this dam-raceway-wheel complex would be known as the Richardson Mill.[37]

After the American Revolution, the problem of poor meadow drainage became especially chronic in Wayland, formerly known as East Sudbury. On

June 10, 1789, "Richard Heard and other subscribers" observed that "a large tract of land, heretofore very valuable, is now being flowed in summer rendered of very little value to the owners." They asked "your Excellency and Honors to issue . . . commissioners, with full power to levy and collect money of the Proprietors of the aforesaid lands for the sole purpose of removing said obstructions." By June 4, 1793, the commissioners realized that the ultimate cause of the drainage problem was the river's natural outlet, so they asked the court to "enlarge the power of said commissioner so as to clear the obstruction in said river down to and over the Fordway (so called) at Billerica." An attempt to excavate the Fordway was made, because time-tarnished and rough-weathered piles of waste rock littered the site when visited by the Joint Committee nearly a century later. In response to this failed effort, the residents of the upstream towns became upset because they were being taxed to support this infrastructure improvement but gained nothing.[38]

During this pre-Revolutionary era, the meadowland farmers were also being disadvantaged by a change in water law that favored the intensive concentrated use of water by industry. Earlier colonial water law had been dominated by the principle "Aqua currit et debet currere, ut currere solebat," or "Water flows and ought to flow as it has customarily flowed." After the Revolution, the balance shifted toward the principle of "adverse use," which roughly means "use it or lose it." This doctrine was favorable to industry relative to earlier agrarian interests because it encouraged the "productive, instrumental use of water" in all forms, but especially streams and rivers. A legal principle called "reasonable use" was invoked to support larger water projects for the public good. In this context, "public" meant the urbanizing, industrializing population beginning to dominate political power.[39]

3 | American Canal

DURING THE FIRST FEW DECADES OF THE NEW REPUBLIC, the lower Merrimack River Valley became an American Ruhr. Its main stem and the principal tributaries—especially the Nashua—were being rapidly developed for timber, farm products, and hydropower. As the crow flies, these places were not that far from Boston, the epicenter of New England's maritime commerce. But getting products to that market required barge transport down the lowermost Merrimack to Newburyport, followed by a sea voyage on the open Atlantic around stormy Cape Ann. It didn't take long for the rising merchant class to realize that a commercial canal cut through the diluvium— glacial sand, gravel, and hardpan—between Boston and the future site of Lowell would save the time and trouble of a sea voyage, thereby raising profits. Though expensive in the short term, it would save money in the long term by increasing efficiency.

Thus plans were made for what would become known as the Middlesex Canal. This project was charted by the Commonwealth in 1793 as a private shareholder corporation, though the legal complexities involved with land acquisitions and water privileges required so many acts of the General Court that it bordered on being a quasi-governmental entity. Though the political momentum to build the canal was strong, details of the enabling legislation were very poorly considered. Writing in hindsight from 1860, Chief Justice Parker of the Massachusetts Supreme Judicial Court opined that this "Act of the Legislature is, indeed, obscure, confused, and almost unintelligible."[1]

The initial plan was to ignore the Concord River as a source of water to plumb the canal. But on March 25, 1794, apparently without any public input, the proprietors of the Middlesex Canal Company purchased the water privilege for the Concord River from the Richardson mill, whose legal claim traced back to Osgood's original gristmill privilege of 1704. Had public use of the Concord River been made public knowledge, the meadowland farmers

FIGURE 12. Factory canal. The Assabet River is channelized in downtown Maynard, Massachusetts to power a historic mill. November 2015.

of the valley almost certainly would have raised the alarm "through every Middlesex village and town" and perhaps been granted an exemption, as was the case for the town of Billerica when the Shawsheen River was being publicly considered.[2]

Purchasing the water led to an act passed by the Commonwealth on February 28, 1795, requiring and "authorizing the Proprietors of the Middlesex Canal to render Concord River boatable" as far upriver as the "Sudbury Causeway" and "to connect Concord River with the Middlesex Canal, for that purpose." "The enterprise was not regarded as threatening to bring water upon the Meadows, but to reduce it by means of the excavations." Meeting these two objectives simultaneously required either blasting a deeper channel at the Fordway or raising the dam. They chose the latter. By 1798 the Richardson dam had been raised a foot higher than the old gristmill and strengthened. By 1803 the canal had been dug and then plumbed by the Concord River, and the weedy sandbars of Musketaquid had been dredged for the passage of large boats called scows. The canal was officially open for business.[3]

Though the objective of the Canal Corporation was very clear—to make money—the end result was a complex piece of infrastructure that didn't profit its shareholders for the first sixteen years after it opened. In its day, it was the longest canal in the nation, crossing the most complex terrain, a twenty-seven-mile-long continuous ditch that was twenty feet wide at the base. It required twenty lock chambers, fifty bridges, and a continuous towpath. Though only marginally profitable for a few decades, it helped serve as a model for the Erie Canal, which sent a delegation to the Middlesex Canal in 1817.

Residents of the meadowland towns had mixed feelings about the canal. Forward-thinking residents saw great possibilities. Transporting a heavy load of wood from the hills south of Framingham, clay and bog iron ore from Pantry Brook in Concord, lime from Carlisle, sacks of grain from Billerica, and bushels of cranberries from Bedford became much easier than before. By 1807 some Concord residents were asking the Canal Corporation to improve this transportation corridor even more by creating a "high road to market" for freight. These improvements would have required raising the dam even more and/or further dredging channel obstructions. By 1835, Concord historian Shattuck was still remarking that "with little expense, it is thought, there might be a profitable inland navigation," even for towns as far upstream as south Sudbury.[4]

ONE OF THOSE WHO TOOK NOTICE of Concord's transition from sleepy river village to industrial waterway was Henry's paternal grandfather, John Thoreau. In 1800 he bought a house that's now part of the Colonial Inn on the north side of the town green, and moved to Concord. His story began in the seaside village of St. Hillier, on the Isle of Jersey in the English Channel, where French Protestants, or Huguenots, settled in 1754 after fleeing religious persecution in Catholic France. Young Jean Thoreau took to the sea as a privateer; after being shipwrecked and rescued, he came to the Massachusetts Bay Colony in 1773, apparently against his will. Making the best of it, he opened a trading shop on Long Wharf in Boston, anglicized his name to John, and became a successful merchant before moving to Concord at the turn of the century. There he raised his son John, who became Henry's father.

THE ORIGINAL PLAN FOR THE MIDDLESEX CANAL called for a deeper excavation and a lower dam. That plan was thwarted by the discovery of serious ledge in the channel. The contractors insisted on at least "three and a half feet upon the rocks in the bottom" to float the canal boats, which were often heavily laden with bog ore for making iron, marble for making lime, brick-making clay, manufactured brick, cordwood, and hay. Unable to go deeper, they raised the Richardson dam. "The water of the River" was thus "thrown back, with the double purpose of making the river boatable and of digging the Canal more cheaply." By 1808 complaints were increasing from upstream farmers that this higher dam slowed drainage, there were more frequent floods, summer floods lasted longer, the ground was softer and wetter, and the quality of hay had deteriorated. According to the 1859 court testimony, by 1809 the canal proprietors had raised the dam a *"foot and a quarter more,* making *two feet and a quarter* in all, being *five inches* more than the entire fall of the River between us and the Dam . . . then they put a sill atop of the crest, and flash boards atop of *that.* So, for the first time, a suit was brought against them by Captain

FIGURE 13.   Dredging Musketaquid. Historic photograph by Alfred Munroe showing the dredging of the Concord River, 1895. Shows steam-fired machine for clearing sediment and weeds from the bars choking the waterways. Courtesy of the Concord Free Public Library.

David Baldwin, a large meadow-owner, on behalf of all. . . . The action was entered into the Common Pleas, April 1810."[5]

In preparation for trial, both sides agreed to have a careful leveling survey done of Musketaquid in 1811 by Col. Loammi Baldwin, the region's most notable civil engineer. According to David Heard, who would later assist Thoreau in his work, the 1811 survey was considered so good and so objective that "both parties relied upon it; and there were no objections to it from any quarter." This case was tried by jury in the Supreme Court during the October term in 1811 with David Heard as a plaintiff. "The verdict," he wrote, "was given against us, and judgment entered up for costs to the defendants." Though the verdict was immediately appealed, the decision "was again rendered for the defendants [the canal corporation], and judgment for costs." These verdicts, alleged the plaintiffs, were attributed to illegalities on the part of the canal "superintendent," who "was openly charged with the artifice by several gentlemen, and answered evasively." Baldwin, a prominent shareholder of the Middlesex Canal Corporation, declared he was "ashamed to belong to such a Corporation."[6]

Frustration from that defeat simmered through 1816, which was New England's "year without summer." The previous year, the violent eruption of Mount Tambora in Indonesia injected a sun-blocking veil of volcanic ash and sulfur into the global stratosphere. In New England, this manifested as a persistent haze or "dry fog" that darkened the sky, caused temperatures to plunge, and intermittently frosted the foliage all summer long. During this unsettled year, farmers to the north experienced complete crop failures. Many packed up and headed west in a largely unheralded migration. In the Concord River Valley, however, the unrest gave birth to the Sudbury Meadow-Owner Corporation, created "for the purposes of clearing out the Concord River, by the removal of the bars, sand, grass, *et cetera*, in the stream."[7]

With complaints against the dam going nowhere, David Baldwin (Loammi's brother) and others decided to work locally: "The meadow-owners then procured an Act incorporating them for the purpose of laying assessments, collecting money, buying machinery, and hiring labor, to scour and deepen the channel. For several years, with huge implements, swimming oxen, and men neck deep in mud and water, we wrought, as best we might, and as we could spare time, in the channel of the River. Thus we added, to the expense of our law-suits and the bills of cost recovered by our adversaries, several

thousand good dollars more. It was all in vain. No removal or sensible miti-
gation of the evil was effected."[8]

River Boyhood

The second son of John Thoreau and Cynthia Dunbar Thoreau was conceived in that year of volcanic and riverine unrest. He was born the following
summer on July 12, 1817. He was baptized three months later in First Parish
Unitarian Church on October 12, where he was given the name David Henry.
His birth town was a lovely, largely agricultural landscape, hugging its three
blue waterways from two sides each, yielding six different river panoramas.

His birthplace was a farmhouse off Virginia Road, east of the town
center, on the alluvial meadows of the Bedford Flats. William Ellery Channing, Henry's trusty sojourning companion and first biographer, gave us this
description: "It was the residence of his grandmother, and a perfect piece of
our New England style of building, with its gray, unpainted boards, its grassy,
unfenced door-yard. The house is somewhat isolate and remote from thor-
oughfares . . . the more smiling for its forked orchards, tumbling walls, and
mossy banks. About the house are pleasant, sunny meadows, deep with their
beds of peat, so cheering with its homely, hearth-like fragrance; and in front
runs a constant stream."[9]

The first decade of Henry's life was geographically unsettled. The family
moved from one place to another as his father's business pursuits teetered on
the edge of bankruptcy: from Concord to Boston, then to Chelmsford, and
then back to Concord in 1823 when John Thoreau was invited to join a pencil-
making business founded by his quirky brother-in-law, Charles Dunbar, and
another local resident. Within a year, the two founders of the improving business left it to John, a company that would provide his main income for the
rest of his life. The year was 1824. Henry was seven. The family consisted of
his mother, Cynthia, his father, John, his older siblings, Helen and John junior, and his little sister, Sophia. Thirty-one years later, Henry recalled one
"evening, with the aid of Mother, the various houses (and towns) in which I
have lived." Up to this point, the list included the "Minott House" and "Red
House" in Concord, an unnamed house in Chelmsford, "Pope's House"
in south Boston, and "Whitwells House" on Pinckney Street, Boston. In
the spring of 1826 the family moved back to Concord to reside in the "Brick
House."[10]

As a boy Thoreau loved water. He didn't cry when baptized, or so he was told. He was considered amphibious, "known among the lads of his age as one who did not fear mud or water." Like all children of the era, he virtually lived outdoors, coming inside mainly to take meals, sleep, and be schooled. A small gang of playful boys included Henry, his brother John, and "two of their schoolmates, the Hosmer brothers from rural Derby's Bridge—Benjamin and Joseph." The common focus of their outdoor life was the Sudbury River between their houses. No doubt the kids had easy access to small boats. Years later, when contemplating the seriousness of adulthood, Henry wrote: "I think that no experience which I have to-day comes up to, or is comparable with, the experiences of my boyhood," when "my life was ecstasy. . . . I can remember that I was all alive, and inhabited my body with inexpressible satisfaction." A large part of that ecstasy involved fishing, shooting, swimming, skating, and boating, mostly along the river. Adult fishermen, he later realized, were those who kept the "pleasures of my earliest youth" without being "confounded by many knowledges." On one occasion the adult Thoreau mistook a swimming boy for a river otter, making it quite likely that someone once thought the same of him.[11]

THE SUMMER OF 1825 WAS SO SCORCHING that fish died in the now more sluggish river. The Erie Canal opened, undercutting the profitability of New England agriculture. The Middlesex Canal, now in desperate financial straits, sold part of the flow of the Concord River on its north side to Francis Faulkner for his "Canal Mills." Thirty-four years later, Faulkner would be one of the defendants in the 1859 flowage controversy. Notably, his deed specified that the Canal Corporation must maintain high water in the reservoir: "The water in the mill-pond when on a still level" must be "at or within three-fourths of an inch of the top of the dam or flash boards as the same now exists on the main dam across the Concord River."[12]

This stipulation for permanent high water was made for a dam too weak to depend on. Having been built of wooden timbers backed up by loose earth, it was then being sorely tested by flood flows. To ensure against future failure, and the lawsuits that would certainly follow, the proprietors of the Middlesex Canal decided to build a bigger, better dam immediately downstream from the existing one, which it would submerge. Unlike its lower, leaky predecessor, this new dam would be "thoroughly cemented," less leaky, and "*two feet and a quarter higher*" than the preceding one; making the whole increased altitude, since the Canal proprietors came into possession, *three feet and three inches.*" Above the top of the dam were flashboards, strong planks eleven inches wide hinged to the top of the stone spillway to back up extra water when most needed in summer.[13]

This new dam was completed in September 1828. During the subtropical storm season the following August, the meadowland farmers living far upstream were "served with notice by a great and protracted flood." "We had a powerful rain, up at Wayland—or Sudbury. . . . That was the 4th, 5th, and 6th of August. In consequence of this rain, we saw that the water did not abate as usual." Worried about this alarming new situation, "some of the Meadow-owners met at a hotel, and a Committee was appointed," and "the meadow proprietors commenced a fresh suit at Common Law, for the unlawful erection and maintenance or heightening and tightening of the dam." During testimony, David Heard wrote: "After 1828, there was quite a change in the stream . . . the water in the River stood higher," especially the "low-water marks . . . where we used to drive our teams, it was impossible to go with them." What "we called a low state of water" then was "two feet higher than previous . . . —say—from 1816."[14]

During the final arguments for the 1859 hearings, "it was thus settled, without controversy, that the Dam of 1828 is 26 inches above the cap-sill of that of 1798. Add, to this, the increased height, of amounting to a foot, of the Dam of 1798 over that which preceded it, and it would be seen that the present Dam exhibits, in all, an addition of 37 or 38 inches to the height of the Dam as it stood prior to 1798." Unfortunately for the farmers, "this trial was lost, the plaintiffs said, because of evidence tampering during a pre-trial experiment." This was the third time that allegations of improper conduct had been made against the industrialists, who were consistently supported by the courts. Following this case, the court decided that the dam owners "had a right to raise the Dam, at any time, and to any height necessary or expedient for the purposes of their creation.[15]

Though the Middlesex Canal proper ran from Lowell to Boston, the stipulation for permanent high water in the dam pool flooded the Fordway deeply enough to allow the passage of canal boats into Musketaquid. There they could travel its "dead" water reaches as far as Sudbury, hauling whatever commodities would pay. "They commonly carry down bricks or wood,"

Thoreau recalled in *A Week,* "fifteen or sixteen thousand bricks, and as many cords of wood, at a time." Iron ore was also a common haul. This was bog iron, nodular rusty masses excavated from places where the geochemical conditions were just right, usually where a strong spring flowed beneath a swamp. One mine on Pantry Brook sent ore to Saugus, the birthplace of the American iron industry. During deluges, spoils from this mine washed into the Sudbury River to concentrate as Robbin's Bar, which would later play a role in the flowage controversy.[16]

THOREAU'S COMPELLING CHILDHOOD MEMORIES of the Middlesex Canal are the earliest, most explicit descriptions of the Concord River documented by his writing. He recalled with wonder the years of barge traffic, when large scows traveled past Concord on voyages to far-off lands. The size of these boats staggered his imagination: "We used to admire unweariedly how their vessel would float, like a huge chip, sustaining so many casks of lime, and thousands of bricks, and such heaps of iron ore, with wheel-barrows aboard,—and that when we stepped on it, it did not yield to the pressure of our feet."[17]

More important, this high-water highway linked his sleepy hometown to the rest of the world. "The news spread like wild fire among us youths, when formerly, once in a year or two, one of these boats came up the Concord River, and was seen stealing mysteriously through the meadows and past the village. It came and departed as silently as a cloud, without noise or dust, and was witnessed by few. . . . Where precisely it came from, or who these men were who knew the rocks and soundings better than we who bathed there, we could never tell. We knew some river's bay only, but they took rivers from end to end. They were a sort of fabulous river-men to us. It was inconceivable by what sort of meditation any mere landsman could hold communication with them."[18]

Reflecting back from 1849, Thoreau recalled being a youth who wanted his river to be truly navigable, so that his town could be permanently linked to the wider world: "We might then say that our river was navigable,—why not? In after years I read in print, with no little satisfaction, that it was thought by some that with a little expense in removing rocks and deepening the channel, 'there might be a profitable inland navigation.' *I* then lived somewhere to tell of."[19]

By the time the Billerica dam was raised in 1828, Thoreau's scholarly tendencies had become self-evident. Pulled from public school at age eleven, he was enrolled in the Concord Academy under the tutelage of Phineas Allen. There he received an education strong enough for him to pass the entrance exams for Harvard College as a sixteen-year-old.

In 1832 meadowland farmers followed through with yet another lawsuit against the dam owners, charging that they had no right to raise the dam beyond what the original charter of the Middlesex Canal allowed. Again they lost. In response, David Heard, acting as "one of the Proprietors of the Meadows Corporation . . . brought three or four suits, afterwards, in another form, one of [Heard's] own and two or three others in other individuals' names." For the culminating trial, a new and even more careful leveling survey of the river was made in 1833 by B. P. Perham under the supervision of Loammi Baldwin, then a civil engineer. During this second prequel to the 1859 case (the first was in 1811), testimony was heard regarding drawdown experiments, which monitored the river's response to preplanned changes in the height of the dam pool. Though none of these many lawsuits succeeded, the last one left an important legacy: "Baldwin's second map," which became the base for Thoreau's scroll map, the centerpiece of his scientific river project.[20]

IN 1833 HENRY ENROLLED IN HARVARD COLLEGE. This was a tough family decision, given the financial stress it would impose on the household budget, and given that Henry's skills with his hands were equal to those with his head. In fact, his parents came close to apprenticing him out as a carpenter. Though he went off to college, Henry learned carpentry on his own, helping to build the family's Texas House, his one-room house at Walden Pond, and his own furniture: geology specimen cases and a chest for his manuscript journals. He also became a capable shipwright, building three boats from scratch before the end of the 1830s. The first was fabricated in 1833 during his final term at the Concord Academy, when he was only sixteen years old. Christened *Rover,* its subsequent history is vague, though this was likely the rowboat he used on Walden Pond in that famous *Walden* metaphor about fate letting him drift to distant shores during the forenoons.[21]

Midway through college, Thoreau became sick with the tuberculosis that eventually killed him. When on leave in Concord, he wrote a reminiscence on April 20, 1835, when he was only seventeen years old. In it he reviews visits with his brother, John, to overlook the Sudbury River from the cliffs of Fairhaven. "Seating ourselves on some rocky platform," they caught "the first ray of the morning sun, as it gleamed upon the smooth, still river, wandering in sullen silence far below." There one could find a "beautiful river at your feet, with its green and sloping banks, fringed with trees and shrubs of every description."[22]

AT THE TIME, THIS HIGHER, deeper river was the southern extension of an industrial canal. Though it remained operational, the water over the gravel bars near the mouth of the Assabet River—particularly Barrett's Bar—was becoming too shallow for commercial traffic. Nathan Barrett, who owned the adjacent land, described how, "in September, 1835, a man from Billerica asked me to have some of my men at work on the bar 'dredging.' They ploughed and scraped out, three or four days. They made the channel on the South side, and it has always remained so. . . . The clearing out in 1835 was for the purpose of enabling boats to go through without grounding. And soon after that, the boating was given up, so that it was not necessary to plough it out again."[23]

This description suggests the following sequence: a river that was easily navigable in the 1820s had become choked with sediment, which required dredging in 1835, and which refilled with sediment after that. Almost certainly the source of this sediment was the spasm of erosion in the lower Assabet associated with the building of the Union Turnpike Bridge in 1827. This was a case of one transportation technology slapping the face of another.

IN 1836 THOREAU WAS HOME ON LEAVE FROM COLLEGE AGAIN. This time he was living a young boatman's life. His first extant piece of writing on this topic was from a letter dated August 5, 1836. It was a nautically themed mini-memoir, written for Charles Wyatt Rice, a friend and classmate. David Henry, as he signed himself then, was building a boat to "keep soul and body together." Specifically, he was "manufacturing a sort of vessel in miniature, not a *eusselmon nea* [well-benched ship] as Homer has it, but a kind of oblong bread-trough." Using his "frail bark," he planned to "cross the purling wave and gain the destin'd port." Thus he would "leave care behind, and drift along our sluggish stream at the mercy of the winds and waves." Within the letter is a long "extract from the log-book of the *Red Jacket*, Captain Thoreau."[24]

After graduating from Harvard College in 1837 with an emphasis on languages and the classics, the twenty-year old Henry returned to Concord to live with his family. By then, they had moved further west on Main Street to the larger Parkman House, located where the Concord Free Public Library is today. That family move brought the river even closer, now just across the street and down the path. As a new graduate, Thoreau commenced a new era in his life by changing his name from David Henry to Henry David. His friends and some family members cooperated; many did not. For the rest of his life some of the townfolk who knew him as a child annoyed him by continuing to call him David. Just as stubbornly, he insisted that he was a new man. The subsequent decade would be a time of great growth, uncertainty, and volatility for Henry. Hopes were defined and dashed. Loves were found and lost.

The salient event of that year was meeting Ralph Waldo Emerson and starting a journal on October 22, 1837, at the elder man's suggestion. Its earliest entries ooze with nautical language. As with his 1835 letter to Charles Rice, they combine his lifelong love of boating with his more recent love of the literary classics. He was especially drawn to Homer's epics of the seafaring Greeks and the Norse sagas. The entry for November 3, 1837, titled "Sailing With and Against the Stream," likely describes leaving his "port" above the T-junction to travel with the stream on the lowermost Sudbury River before heading against the stream on the Assabet. Another entry that same week, "Still Streams Runs Deepest," dates to November 9, 1837, and describes the lakelike Sudbury reach of Musketaquid. In this reach, the river had recently been deepened and broadened by the growth of a gravel bar at the mouth of the Assabet that had been elevated by some combination of a higher Billerica dam and enhanced sedimentation.[25]

Henry's earliest seagoing voyage was on May 3, 1838. Having just quit his excellent teaching post in Concord because he refused to use corporal punishment on his students, he was en route to Maine in search of another teaching job. After steaming out of Boston and rounding Cape Ann, he reached the open sea, got seasick, and paid his "small tribute to Neptune." Not wanting to miss the experience, he stayed on deck late into the night with his "head over the boat's side—between sleeping and waking." Beyond Portland, he sailed on the *Cinderella* to Castine in Penobscot Bay on May 13. Four days later he shipped back to Boston.[26]

Thoreau returned to Concord in failure, having not secured a teaching position. After a brief transition period, he opened a private school with his brother John that lasted just a few years. His journal often dealt with the three rivers. The entry for September 16, 1838, contains an omen of his future involvement in the flowage controversy, which was then twenty-one years in the future. It contains his first complaint about damming rivers: "Dam it up you may, but dry it up you may not, for you cannot reach its source. If you stop up this avenue or that, anon it will come gurgling out where you least expected."[27]

"The Fisher's Son," his poem dated August 13, 1838, and "Rivers," dated September 5, are both infused with nautical jargon. His ode "Fair Haven," from December 15, 1838, exhibits his love for a place that combined the advantages of open water and easy access via a navigable stream. This place was so special that Henry gave it six other names: Fair Haven Bay, Fair Haven Pond, Fair Haven Lake, Fair Haven, Fairhaven, and simply "the pond," a name he would later apply to Walden Pond. By the end of his first year out of college he had penned one of his most durable nautical metaphors about the river, which would enter the published text of *Walden* sixteen years later: "I return again to my shoreless—islandless ocean and fathom unceasingly for a bottom that will hold an anchor, that it may not drag."[28]

During the spring of 1839, Henry built his third boat, this time with the assistance of his brother, John, and with the purpose of using it for teaching trips. This was the only boat of his life that he described clearly: "Our boat was built like a fisherman's dory—with thole pins for four oars. Below it was green with a border of blue. . . . It was well calculated for service—but of consequence difficult to be dragged over shoal places or carried round falls. . . . Two masts we had provided, one to serve for a tent pole at night, and likewise other slender poles that we might exchange the tedium of rowing for poling in shallow reaches . . . a tent of drilled cotton—eight feet high and as many in diameter." Later he gives the dimensions, a "dory—15 feet long by 3 in breadth at the widest part—a little forward of the center." Eventually he added a set of rollers, or wheels, for portaging this heavy boat around rapids. And the cost? "A week's labor" the previous spring. He named it the *Musketaquid,* after the alluvial valley he voyaged in.[29]

During the halcyon days of teaching with his brother, they made frequent trips up the lower Assabet, which was more private than the other two river spokes, and more strongly flowing. These trips inspired Henry's first long ode to river travel, sixty lines of descriptive verse that entered his journal on July 18, 1839, as "The Assabet." This was six weeks before their more famous trip up the Merrimack, and may have helped give him the idea. This poem predates his understanding of the dramatic changes taking place on the Assabet due to human impact, but not his recognition of their effects: the rapids, foaming water, yellow sand, shallow bars, pebbly bottom, tipped trees, and rocks in the channel.[30]

ON AUGUST 31, 1839, Henry and his brother left town for what he called their "White Mountain expedition." He claimed at departure to have "weighed anchor" when, in fact, they likely unchained their dory from the same bankside apple tree they would return it to. For this trip, the Merrimack was mainly a means to an end: an inexpensive and pleasant mode of travel giving them access to New England's highest peaks. Though the vacation lasted two weeks, only one was spent on the water, and all of that was effectively industrial water. The "deep, dark, and dead stream" they headed down resembled "a long woodland lake bordered by willows," and for good reason: it had been rendered more canal-like by the Billerica dam. After the short, flowing reach near the Fordway, they entered the pool of the dam. From there they "entered the canal, which runs, or rather is conducted, six miles through the woods to the Merrimack at Middlesex." With no slope to the water surface and no wind, they pulled their boat along the towpath under human power. Given the excavated channel, Thoreau found "some want in the scenery." When passing through it, he looked forward to the time when it would be indemnified by natural decay. Marking the end of the towpath was the canal lock that lowered boats down to the level of the Merrimack, just above Pawtucket Falls in Lowell, the site of the largest factory complex on the continent.[31]

During the year that Henry and John boated down the Concord River, the high water that made their trip possible was being challenged in court. The Supreme Judicial Court of Massachusetts had recently established a special adjudication procedure for claims against dam owners and canal operators who caused flowage on private lands. Taking advantage of this new legislation, which allowed more than one year after a change to make claims, the meadowland owners "commenced a fresh suit at Common Law" against the

Middlesex Canal for the years 1799, one year after the dam was built, and 1829, one year after it was raised. Their case was dismissed.[32]

FOR THE NEXT THREE YEARS, between 1840 and April 1843, Henry's journal over-flows with nautical and fluvial glosses. In February 1840, his poem "The Freshet" reveals his prescient gift for observing the details of river hydrology; later he would quantify the rise and fall and timings of flood waves with mon-itoring data. He would also make use of similar language on many occasions, referring to the daily tide, steering by the points of the wind, finding a good anchorage for friendship, treating one's thoughts as if they were vessels that sail away, praying to be "at the helm at least once a day" so that we might "feel the tiller rope in our hands," identifying with Robinson Crusoe, commit-ting oneself to sailing through life by a "sort of dead reckoning," imagining heaven as a place of anchorage, aspiring to "sail away from winter." These are just a few examples among dozens demonstrating that Thoreau's identity as a boatman preceded and greatly influenced his early literary career long before he decided to move to Walden Pond.[33]

More examples: "I sailed from Fair Haven last evening as gently and steadily as the clouds sail through the atmosphere. The wind came . . . like a winged horse. I could watch the motions of a sail forever, they are so rich and full of meaning. I watch the play of its pulse as if it were my own blood beating there. . . . So am I blown on by God's breath, so flutter and flap, and fill gently out with the breeze." The "calendar of the ebbs and flows of the soul; and on these sheets as a beach, the waves may cast up pearls and seaweed." "He is the best sailor who can . . . fill his sails longest," and who "extracts a motive power out of his obstacles."[34]

Were it not for the waters of the river and Fair Haven Bay, he wrote in December 1840, he would "wither and dry up," much as "the muskrat or the herbage on their brink." "My life will wait for nobody," he wrote on April 7, 1841. "It will cut its own channel—like the mountain stream which by the lon-gest ridges . . . is not kept from the sea finally. So flows a man's life." Such a man could never live away from the water for long, not even when he decided to become a farmer-landowner. In Henry's case, this happened in April 1841. Unsettled by the recent closing of his school and with no prospect of a profes-sional income, he purchased the riverfront Hallowell farm and began to gather some farming equipment. However, the farm's owner changed his mind and wanted it back. Thoreau agreed to drop the deal. Two years later, he returned once again to the idea of buying a farm, but again, only on the condition that it be "bounded by a river. . . . It would increase my sense of security and my energy and buoyancy when I would take any step."[35]

In the autumn of 1840 Thoreau credited the Concord River with being "the stream of our life," an "emblem of all progress, following the same law with the system, with time, and all that is made," and the "direction of fate." Rivers and human lives are never stationary, but always flow from one time and place to another; trees are but "rivers of sap," the heavens a "river of stars," and rocks "rivers of ore." In all these quotes, there is enthusiasm for the continuous flow of matter and energy that is distinctly absent from his later descriptions of Walden Pond, with its invisible groundwater seepage.[36]

THE WRITING OF A WEEK BEGAN IN SEPTEMBER 1839 with notes scribbled in a notebook during the trip with his brother. In 1840 these jottings were expanded and organized, and then set aside once again. In early 1842 tragedy struck the Thoreau household: John junior died unexpectedly and painfully from lockjaw. Henry's initial grief response was psychosomatic: a replay of the symptoms of his brother's fatal illness. Later he channeled that grief into a plan to memorialize his brother with a book-length account of their trip together. Though Henry attempted to write this book in 1842, 1843, and 1844, the vicissitudes of his unsettled life prevented this from happening.

Within a few months of John's death, a young writer named Nathaniel Hawthorne and his new bride, Sophia, came to Concord to live. They decided to rent the Old Manse, which had waterfront views of the Concord River just south of the site of the Old North Bridge. Waiting for their arrival on July 8, 1842, was a new garden planted by Henry Thoreau, who had been hired for the task by Ralph Waldo Emerson. On August 31, Henry invited Hawthorne for a row in *Musketaquid,* after which he sold this boat to his new friend. They soon became good friends, taking many river excursions together, mostly up the Assabet.

When Hawthorne departed for Salem four years later, in 1846, he bequeathed his boat—renamed the *Pond Lily*—to William Ellery Channing, under whose care it eventually rotted into oblivion. At that time, Thoreau was living in his one-room house at Walden Pond, voyaging the "Atlantic and Pacific" of his own consciousness. There he was using an old boat that

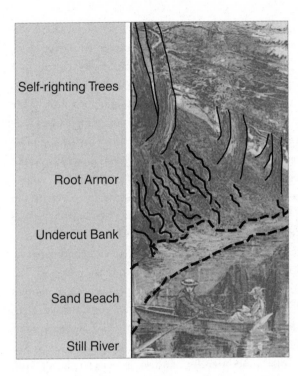

FIGURE 14. Picturesque hemlocks. Highlights and labels overlain on semi-transparent illustration from *Concord Guide Book 1880* by George Bradford Bartlett, drawn by L. B. Humphrey and Robert Lewis. Illustration shows regrowth of trees on bank that stabilized after Thoreau's era.

he apparently found on Walden's shore. Before leaving Walden Pond in September 1847, Thoreau had already written a full draft of *A Week,* apparently modeled after Margaret Fuller's *Summer on the Lake.* Because this book about his 1839 river trip took a decade to publish, much of its lyrical riverine writing comes from Thoreau's later, more refined observations from the 1840s.[37]

During the summer he befriended Hawthorne, Thoreau published his first piece in the genre that would make him famous. At the invitation of Ralph Waldo Emerson, he published "Natural History of Massachusetts" in the July 1842 volume of the *Dial* to celebrate publication of a scientific survey sponsored by the Commonwealth. Specifically, Emerson had "begged our friend to lay down the oar and fishing line, which none can handle better, and assume the pen"—another independent early testimonial to Henry's skill as a boatman. In the essay, Thoreau compares Concord's "broad lagoons" and "yonder fen" to Venice, making it easy to imagine him propelling his boat as a one-oared gondola through slack waters. He also expresses concern for the fate of the town's bridges owing to the destructive force of ice floes. This is the topic that would launch his river project seventeen years later.[38]

Thoreau's second early essay on natural history, "A Winter Walk," was published in the October 1843 issue of the *Dial*. This was mainly a river walk, accomplished with what he called his "swift shoes," his ice skates. For the first time Thoreau described the river as a network of gently inclined lines sloping up from the flatness of its main stem to the steepness of its highest rills. This became a recurrent theme in his life: the river as a pedestrian highway, especially during winter. "We may go far up within the country now by the most retired and level road, never climbing a hill, but by broad levels ascending to the upland meadows. It is a beautiful illustration of the law of obedience, the flow of a river. . . . Its slight occasional falls, whose precipices would not diversify the landscape, are celebrated by mist and spray, and attract the traveller from far and near. From the remote interior, its current conducts him by broad and easy steps, or by one gentle inclined plain [*sic*], to the sea. Thus by an early and constant yielding to the inequalities of the ground, it secures itself the easiest passage."[39]

Also within this essay are wonderful early descriptions of what sounds like the Sudbury River: "This meandering river . . . following the winding of the stream, now flowing amid hills, now spreading out into fair meadows, and forming a myriad coves and bays where the pine and hemlock overarch . . . the empire of fishes . . . unfathomed depths," and so on. Already he's giving readers a clue to what he would later emphasize: that his path away from a surfeit of civilization was the river: "The river flows in the rear of the towns, and we see all things from a new and wilder side. The fields and gardens come down to it with a frankness, and freedom from pretension, which they do not wear on the highway. It is the outside and edge of the earth."[40]

A third early essay, "Paradise (to Be) Regained," was published in November 1843. As with "Natural History of Massachusetts," it was ostensibly a book review, although it brims with unrelated ephemerae, not the least of which is a veiled criticism of the taming of rivers and brooks by dams: "Men having discovered the power of falling water, which after all is comparatively slight, how eagerly do they seek out and improve these *privileges?* Let a difference of but a few feet in level be discovered on some stream near a populous town, some slight occasions for gravity to act, and the whole economy of the neighborhood is changed at once."[41]

Curiously, this passage simultaneously italicizes and queries the word "privileges." Writing only a few months before threatening sabotage against the Billerica dam, he is questioning whether humans have the right to put rivers in shackles.

4 | Transition

BY LATE WINTER 1844, Thoreau was in a deep funk. His newfound tribe, the transcendental coterie, was unwinding. Its leader, Ralph Waldo Emerson, was disappointed with his young protégé, admitting in his private journal that "H will never be a writer." This would be the spring when Thoreau and his young companion Edward S. Hoar accidentally set fire to the woods while boating the Sudbury River. On that dry windy day in April, the rapidly moving fire threatened the town center with a fiery conflagration and caused significant economic losses to private landowners. To his townsmen, Thoreau became a persona non grata, a "dammed rascal." Robert Sattelmeyer, the editor of the Princeton edition of the *Journal* for this time period, noted that "no dated journal entries after January 7 survive for 1844, and Thoreau published nothing that year after the last issue of the *Dial* went to press. He delivered no lectures, and even his correspondence waned: only three letters from the year are extant."[1]

Thoreau's youthful delight in the navigable high waters of the Middlesex Canal from the 1820s had turned to seething anger. In his undated journal entry from sometime after August 1, 1844, he threatened sabotage against the dam that made the canal possible: "And—who knows—what may avail a crow bar against that Billerica Dam!" A crowbar, of course, is a tool for moving heavy objects small distances—in this case large blocks of stone. His idea was to wreck the dam so that the river could flow freely again. Six years later, he published this sentence nearly verbatim in *A Week*, with one important change: he ended it not with an exclamation point but with a question mark. This change converted a private threat of sabotage into a public question about whether sabotage might be justified. This punctuation sequence is consistent with his maturation: the emphatic exclamation point of his late twenties becomes the not-so-sure question mark of his early thirties.[2]

In 1844 Henry's main sympathies were with the American shad, often called "America's Founding Fish," whose annual migrations up and down

FIGURE 15.   Causeway and bridge. Historic photograph by Herbert Wendell
Gleason. Lee's Bridge, island and flood, from top of Williams House, 1901. Note
great constriction of flow between bridge abutments. 1901.18. Courtesy of the Concord Free
Public Library.

the rivers of New England provided food and fertilizer for the early settlers. "Salmon—Shad—& Alewives—were formerly abundant in this river and taken in weirs by the Indians, and afterward by the settlers who used them as food and as manure—until the dam and afterward the canal at Billerica, and the factories at Lowell put an end to their migrations hitherward. . . . Perchance after a few thousands of years if the fishes will be patient . . . nature will have leveled the Billerica dam and the Lowell factories—and the 'grass-ground' river will run clear again." When he published this thought in 1849, he extended the idea of dam removal all the way up to the headwaters of the Assabet and the Sudbury, asking that the fish be allowed to explore "new migratory shoals, even as far as the Hopkinton pond and Westborough swamp."[3]

"I for one am with thee," he writes of the shad, declaring his solidarity just before threatening to tear the dam down. At this stage of his life, he wanted it gone, along with the whole of the industrial infrastructure associated with the Middlesex Canal, where he found "some want of harmony in its scenery, since it was not of equal date with the woods and meadows through which it is led." In this passage, he laments the change from the previous geological epoch to his own, hoping that "in the lapse of ages, Nature will recover and indemnify herself." His highest hope is that this piece of industrial infrastructure will, given "the conciliatory influence of time on land and water, . . . pass directly out of the hands of the human architect into the hands of Nature, to be perfected." This thinking exemplifies his early dualistic thinking toward nature, his "trouble with wilderness." At this stage he's still caught up in the paradox that his human epoch is a blemish on an otherwise pristine planet.[4]

The shad's liberty was mainly an ethical issue involving piscine animal rights. The much more important economic issue of his day involved the way the Billerica dam was wrecking the upstream meadows, or so he thought then. In an accurate foreshadow of the flowage controversy to come, Thoreau described its impact on Wayland: "This town is the greatest loser by the flood. Its farmers tell me that thousands of acres are flooded now, since the dams have been erected, where they remember to have seen the white honeysuckle or clover growing once, and they could go dry with shoes only in summer. Now there is nothing but blue-joint and sedge and cut-grass there, standing in water all the year round. For a long time, they made the most of the driest season to get their hay, working sometimes till nine o'clock at night, sedulously paring with their scythes in the twilight round the hummocks left by the ice; but now it is not worth the getting, when they can come at it, and they look sadly round to their wood-lots and upland as a last resource."[5]

For proof of the link between the Billerica dam and Wayland's flooded meadows, Thoreau chronicled a specific case, likely from the mid-1840s, when the farmers were surprised by high water after the dam's wooden flashboards (here called floatboards) were installed. "One year, as I learn, not long ago, the farmers standing ready to drive their teams afield as usual, the water gave no signs of falling; without new attraction in the heavens [tide], without freshet or visible cause, still standing stagnant at an unprecedented height. All hydrometers were at fault. . . . But speedy emissaries revealed the unnatural secret, in the new float-board, wholly a foot in width, added to their already too high privileges by the dam proprietors."[6]

Next he laments the long-term suffering of those farming the upper Sudbury River: "At length it would seem that the interests, not of the fishes only, but of the men of Wayland, of Sudbury, of Concord, demand the leveling of that dam. Innumerable acres of meadow are waiting to be made dry land, wild native grass to give place to English [hay]. . . . So many sources of wealth inaccessible. They rate the loss hereby incurred in the single town of Wayland alone as equal to the expense of keeping a hundred yoke of oxen the year round." In this 1849 text from *A Week* Thoreau accurately prophesized what took place during the Joint Committee hearings in 1859.[7]

Thoreau's sympathy for the meadowland farmers remained a consistent theme to the end of his life. They were, after all, the true descendants of the original Puritan settlers, who Thoreau believed were heroes as valiant as those Homer wrote about and philosophers equal to those mentioned by Virgil and Cato. His praise for the Anthropocene conversion is adulatory: "And thus he plants a town. He rudely bridged the stream, and drove his team afield into the river meadows, cut the wild grass, and laid bare the homes of beaver, otter, muskrat, and with the whetting of his scythe scared off the deer and bear. He sets up a mill, and fields of English grain sprang in the virgin soil."[8]

Thoreau went out of his way to acknowledge and appreciate the hard work it took to create the settled landscape he loved: "Nothing has got built without labor. Past generations have spent their blood & strength for us. They have cleared the land—built roads &c for us. In all fields men have laid down their lives for us. Men are industrious as ants." His Musketaquid farmers were the true salts of the earth. Each spring planting was a sacrament that helped restore Thoreau's "long-lost confidence in the earth." "What noble work is plowing, with the broad and solid earth for material, the ox for fellow-laborer, and the simple but efficient plow for tool!"[9]

Each summer he rhapsodized about the communal "vintage" of the haymakers. He valorized the work as "a thirteenth labor" that "methinks would have broken the back of Hercules, would have given him a memorable sweat. . . . To shave all the fields and meadows of New England clean. . . . Mexico was won with less exertion and less true valor than are required to do one season's haying in New England. . . . This haying is no work for marines, . . . troops. . . . It would wilt them, and they would desert. . . . Early and late the farmer has gone forth with his formidable scythe, weapon of time, Time's weapon, and fought the ground inch by inch."[10]

When sailing and rowing past these work parties, Thoreau painted word pictures of their annual event. "I passed as many as sixty or a hundred men thus at work to-day. . . . I hear their scythes cronching the coarse weeds by the river's brink as I row near. The horse or oxen stand near at hand in the shade on the firm land, waiting to draw home a load anon. I see a platoon of three or four mowers, one behind the other, diagonally advancing with regular sweeps across the broad meadow and ever and anon standing to whet their scythes. . . . In one place I see one sturdy mower stretched on the ground amid his oxen in the shade of an oak, trying to sleep." Later he wrote, "There are . . . squads of half a dozen far and near, revealed by their white shirts. . . . A great part of the farmers of Concord are now in the meadows, and toward night great loads of hay are seen rolling slowly along the river's bank,—on the firmer ground there,—and perhaps fording the stream itself, toward the distant barn, followed by a troop of tired haymakers."[11]

After the scything, he saw "the haymakers now raking with hand or horserakes into long rows or loading, one on the load placing it and treading it down, while others fork it up to him." After the stacking, he described and sketched the change in scenery: "Counted twenty haycocks in the great meadow, on staddles [platforms], of various forms—tied round with hay ropes. They are picturesque objects in the meadow." Yet when the job was done, "I hear no boasting, no firing of guns nor ringing of bells." Rather, the farmer "celebrates it by going about the work he had postponed 'till after haying'!"[12]

Diametrically opposed to his support for meadowland farming was his scorn for capitalist market farming on the deforested uplands. Driven by Concord's midcentury market revolution, farmers turned away from farming as a way of life toward farming as a mode of production enhanced by horse-drawn machinery and agricultural chemistry. The soil became less a home than a commodity that could be used to produce another marketable commodity to sell at a profit, mainly in Boston. Consequently, the farm economy of Thoreau's era was being greatly ramped up in scale. In *Walden* we read: "I see young men, my townsmen, whose misfortune it is to have inherited farms, houses, barns, cattle, and farming tools; for these are more easily acquired than got rid of. . . . Who made them serfs of the soil? . . . How many a poor immortal soul have I met well-nigh crushed and smothered under its load, creeping down the road of life, pushing before it a barn seventy-five feet by forty, its Augean stables never cleansed, and one hundred acres of land, tillage, mowing, pasture, and woodlot!"[13]

The eighth labor of Hercules had been to muck out the enormous Augean stables, the huge barns where cattle, pigs, and sheep were slaughtered for export. These were broadly equivalent to the colossal feedlots and factory slaughterhouses of today. What Henry strenuously objected to was the mantra that "bigger is better." Upland farming, he decided, was raining ruin down onto the alluvial landscape and, worse, corrupting men's souls.

Tiger on the Shore

During the summer of 1844, the City of Boston acquired the rights to Long Pond in Framingham for use as a water supply reservoir. By building a dam at the pond's west end, they diverted toward Boston millions of gallons of water that originally flowed to the Sudbury River. During construction of a system of dams, aqueducts, and canals, marketing-savvy developers rebranded Long Pond as Lake Cochituate to make it seem bigger and cleaner. This Native American name conveyed a sense of primitive purity for those who didn't know its actual meaning, "Place of Falling Water"; the colonial English had renamed it "Rocky Falls," "Falls of the Sudbury," and "Saxonville Falls" in historical sequence. This would become the industrial center of urban Framingham, now largely underneath pavement. The 1844 diversion of the upper Sudbury River to Boston via Lake Cochituate was a preview of what would happen nearly a century later in 1931 when the water of the Swift River in central Massachusetts was diverted to Boston via the Quabbin Reservoir, the city's present water supply.[14]

Initially the meadowland farmers were pleased with this idea of having part of the Sudbury River diverted to Boston because they believed it would help reduce the summer flow by about one-third, thereby making the meadows drier. This seems to have been the case, at least for a few years.

During court testimony, John Simonds of Bedford recalled that "in 1845 or '46, the Meadows were very good." Of course, diverting flow to Boston reduced the water discharge to the factories below the Billerica dam. So, as a precondition of diverting the Sudbury, the City of Boston was required to compensate factory owners on the lower Concord River for their loss of power. They did so by constructing two enormous reservoirs to hold winter water for summer release, one each near the headwaters of the Assabet and Sudbury Rivers.[15]

Fifteen years later, the courtroom testimony of meadowland farmers described this 1844 agreement as a "fatal calamity" because it required "two vast Reservoirs . . . one of them is on a branch of Sudbury River, called Whitehall River, in Hopkinton, and occupies an area of five or six hundred acres. The other occupies about three hundred acres, and is situated in Marlborough, on a branch of Assabet River." Farmers complained that during their haying season, July and August, waters from these reservoirs "pour into our valley every day for these months, upward of seventeen millions of gallons," whereas the natural flow from the lake is only "five millions daily." These "deluging powers," these "water avalanches," they claimed, made the meadows much wetter. "Thus our situation is like that of the wretch pursued by an alligator in the water, and confronted by a tiger on the shore, only there is no chance, in our case, that the monsters may jump into each other's jaws!"[16]

Excursion to Walden

During that same summer of 1844—while Thoreau threatened sabotage, and the City of Boston built Lake Cochituate—the Boston and Fitchburg railroad opened for business in Concord. In June, America's first inland river town became linked to the rapidly expanding network of rail opening up the heartland of rural New England, especially the valleys of Vermont and New Hampshire. In the Musketaquid towns, local public attention shifted away from the Concord River as a mode of transport and toward the faster track of the iron horse. Instantly the railroad began to undercut the Middlesex Canal as a means of shipping freight to and from Boston. Within six years the canal would be out of business, the ore boats and barges of Thoreau's boyhood gone forever, and the route of his 1839 journey on the Concord and Merrimack Rivers no longer possible.

As with the case of every settlement farther west, a new railroad opened up new land for development. Speculators moved in for real estate bargains. Just west of Concord Station—on the proverbial other side of the tracks—was a plot of land that opened up for inexpensive house lots. Cynthia Thoreau, tired of either renting from landlords or living as a guest of relatives, seized the opportunity, claimed a lot, and convinced her laconic husband to create a home of their own. This would be known as the Texas House, named for the access street, which itself had been named for the recent annexation of Texas. Henry strenuously objected to this political imperialism.[17]

After buying the lot on September 9, John Thoreau immediately mortgaged the house to pay for materials to build it with. Henry and his father then went to work, completing the house in time for winter habitation. This was Henry's first hands-on, trial-and-error (they forgot the stairs) house-building experience. Five family members would live there, along with whatever boarders and guests they took in. Architectural historian Barksdale Maynard discovered "an old photograph" showing that the Texas House "had a wing about the size and shape of the later Walden house, with its own chimney." Thus this wing seems to have been both the prototype and training ground for Thoreau's one-room house at Walden Pond (yet another reason not to call the structure he built at the pond a hut, cabin, or shanty).[18]

The arrival of the railroad also spiked the price of wood needed to fuel the locomotive. Lands near the tracks, especially Walden Woods, would rapidly be clear-cut for cordwood. Fearing that his favorite walking retreat would also fall to the axe, Ralph Waldo Emerson bought the Wyman Lot on the north side of Walden Pond on September 21. It was an irregular, eleven-acre lot purchased for only $8.10 per acre. By week's end he had added two more acres. Though Emerson described the purchase as an "absurdity," this was disingenuous. By owning the land, he could either keep it as a sylvan walking preserve or sell its trees at a profit when the price was high enough. Emerson was already on his way to becoming a land baron, one of the wealthiest men in town.[19]

By the following March, all three of these railroad stories—the train, the Texas House, and Emerson's purchase—intersected to send Thoreau to Walden Pond, rather than to some other place, for his experiment in deliberate living. The tracks provided a straight, flat, breezy, private, elevated walking path between the crowded family house in town and an empty, available building lot at the pond. Henry could be a free tenant on Emerson's land while simultaneously staying in touch with his family and town friends. The trip, being slightly over a mile, would have taken Thoreau less than twenty minutes, provided he didn't stop to enjoy the symbolic beauty of flowing sand on the banks of deep excavations. So in March 1845 he famously borrowed an axe, walked the tracks to Emerson's woodlot, and begin the process of framing his small house at Walden Pond with "arrowy pines." Ironically, this plan to live at Walden Pond coincided with the completion and delivery of his first-ever lecture on the Concord River.[20]

Thoreau's world-famous move to Walden Pond on July 4, 1845, is linked to the Concord River in several other ways. Producing *A Week on the Concord and Merrimack Rivers* was his highest writing priority in going there. Within two years, two months, and two days, he had written two full manuscript drafts of *A Week*, the first draft of *Walden*, a draft of an account of his travels to Maine that was long enough to be serialized in five installments, and hundreds of pages of other lectures and manuscripts, most famously "On the Resistance to Civil Government," later retitled "Civil Disobedience." Indeed, "Thoreau's move to Walden Pond," noted Robert Sattelmeyer, "was one of a series of temporary living and working arrangements he tried during the 1840s in order to secure time to write." *A Week* was by far the largest of his manuscripts. As early as the spring of 1846, Thoreau was calling it his "big book," being four times longer than *Walden*'s first draft. That summer he read portions of *A Week* to his friends Channing, Bronson Alcott, and Emerson, all of whom encouraged him to proceed.[21]

Henry's interest in the river seems to have intensified while living at Walden. Perhaps this was because he was further away from it than ever before in his Concord life. His river excursions for those two years, two months, and two days became the sources of many of *Walden*'s most memorable phrases. A place without "Neva marshes" and where there was no need to "build on piles" was a counterpoint to the many causeways built across the soft river muck. "A surveyor if not of higher ways than of forest paths" referred mostly to river trails. And, most important, "This is our lake country." This famous phrase follows a description ending with Fairhaven Bay on the river, a place he was then calling "west fair Haven lake."[22]

For Thoreau, Walden Pond was an island of water surrounded by pine and oak woods on the dry plateau of a glacial-age riverbed. Yet when living there, he dreamed of inhabiting an alluvial island surrounded by river water. "I have a fancy for building my hut on one," he wrote, especially one born as the "offspring of the junction of two rivers, whose currents bring down and deposit their respective sands in the eddy at their confluence, as it were the womb of a continent."[23]

With the first draft of *A Week* completed by August 31, 1846, Henry left for a trip to the interior wilderness of Maine. Aside from the climb of Mt. Katahdin, this excursion took place mainly on the rivers and lakes of interior Maine, about which Thoreau wrote extensively. My favorite parts of this work—published posthumously in the Katahdin section of *The Maine Woods*—are his descriptions of paddling rivers in the remote wilderness, the construction of a *bateau* and its handling in whitewater, and the joy of trout fishing in crystalline streams. This excursion, written up as a series of essays in 1847, finally gave Thoreau the first-person material that lecture audiences were eager to hear about and that publishers were willing to pay for. Thus he expanded his lecture repertoire and traveled widely, from Maine to Salem, Newburyport, Worcester, Boston, Concord, New York, and points in between. Between July and November of the following year, he serialized this work in five installments in *Sartain's Union Magazine* under the title "Ktaadn and the Maine Woods." Though the pay for these freelance pieces was a trivial part of his income, Henry had finally become a professional writer.[24]

BY MARCH 12, 1847, Henry had submitted *A Week* to several publishers. Over the next several months, he would learn that none was willing to take the risk. In September 1847 Thoreau left Walden Pond to live at Ralph W. Emerson's house while the philosopher was away in Europe on a lecture tour. Upon Emerson's return to Concord, Henry moved back to the family's Texas House on July 30, 1848. To pay for his upkeep there, he did odd jobs, explored land surveying as a possible vocation, and redesigned the lead-grinding machinery of the family pencil business. During this interval, a four-day walking trip to southern New Hampshire rekindled his interest in *A Week*, especially sites associated with the Indian wars.

In February 1848, with four rejections in hand, he offered *A Week* to another Boston publisher, Ticknor and Company, along with an early version of *Walden*. Though Ticknor agreed to publish *Walden* if Henry guaranteed the costs, they refused to publish *A Week*. They did agree to print it at Henry's expense, which would come to $450 for 1,000 copies, half of them bound and the other half as loose sheets. Being short of cash, Henry returned to James Monroe and Company, which had previously rejected *A Week* but finally agreed to publish it on the condition that its author guarantee the full cost. Though Henry was discouraged from taking this risk by his family, he went ahead anyway, signing a deal that would indebt him for years.[25]

A Week was finally published on May 30, 1849, bearing an advertisement for the upcoming *Walden.* Seventy-five review copies were sent out to stimulate sales. Reviews were lukewarm, and only about two hundred people parted with their money for it. The unsuccessful launch of this memorial to one dead sibling coincided with the loss of another: the death of Henry's older sister, Helen. Four years later, while still in debt for financing the book with a loan, the bulk of the print run of *A Week*—an unsold inventory of 706 copies—was removed from the basement storeroom of his publisher, packed into a wagon, and delivered to the Thoreau house at 255 Main Street, whereupon Henry hauled them upstairs to his garret. The loss was $290, which exceeded his earnings from all four Maine essays.[26]

A Week might have been a commercial success had its author written about the travel adventure that actually took place: two young men in the prime of their lives traveling to the high country of New England to trek above tree line and climb its highest peaks. Instead, Thoreau used the trip as a thinly veiled literary device for a random collection of scholarly essays, poems, and digressions gathered as a plaintive memorial to a brother who had died eight years before publication. Within it are poems that even positive reviewers would have preferred to see excised. Seven pages were reserved for fish zoology. The diatribe against Christianity was relentless. The account of the final ten miles of the homeward trip includes an extended essay on Chaucer. There's a primer on Oriental philosophy. There's a description of a cross-country ramble that Thoreau took to Mount Greylock, in northwestern Massachusetts, in 1844, nearly five years after the boat trip and two and a half years after his brother's death. There are heartfelt confessions of unrequited love for Ellen Sewall, to whom he proposed marriage in November 1840 and who remained friends with him the rest of his life. There are passages about deep time that closely paraphrase the thinking of geologist Charles Lyell without attribution.[27]

Notwithstanding his first book's utter commercial failure, Henry committed to being a writer. He declared this publicly in a letter to Harvard College's president, Jared Sparks, to justify his need for library privileges, despite not meeting the qualifications of being a clergyman and living within the arbitrary ten-mile radius. "I wish to get permission to take books from the

College library to Concord, where I reside . . . because . . . I have chosen letters for my profession, and so am one of the clergy embraced by the spirit at least of the rule," he wrote. "Moreover, though books are to some extent my stock and tools, I have not the usual means with which to purchase them."[28]

WHEN HENRY LEFT WALDEN POND IN SEPTEMBER 1847 to live and work at Emerson's house, he had limited access to Concord's rivers. Perhaps this helps explain why, during this interval, he traveled frequently, usually by railroad. On October 9, 1849, Henry left by train with Ellery Channing for his first trip to Cape Cod. Though they were planning to take a steamer to Provincetown, they discovered that the vessel had been delayed by a heavy storm—the same storm that had sunk the brig *St. John*. Henry and Channing decided to continue their trip via land instead and stopped in Cohasset, near where the sinking had occurred. Henry's description of the wreckage, with bodies floating ashore, would become the poignant opening and the most philosophical section of his first Cape Cod essay. From Cohasset they took a stage to Bridgewater, where they caught a train to Sandwich. From there they took another stage to the elbow of Cape Cod, and then walked from Eastham all the way to Provincetown. The steamer *Naushon* returned them to Boston, whereupon they rode the train back to Concord. Henry returned to Cape Cod alone in 1850 to collect additional material for several paid travel pieces about Cape Cod, later gathered into a posthumous book.

In July 1850 Henry made a solemn trip to Fire Island, New York. He had learned of the wreck of the *Elizabeth,* a ship that was bringing his friend Margaret Fuller, a brilliant intellectual and leading feminist, back from Europe, and went to search for her remains, which were never found. At one point there had been rumors that Henry and Margaret were an item, perhaps because they were transcendentalist colleagues and had stayed simultaneously at Emerson's house, but Henry had quickly disavowed that speculation.

Two months later, on September 25, 1850, Henry and Channing signed up for a package tour to see Quebec. This was a promotional event sponsored by the railroad, which was offering steeply discounted fares. They traveled by train to Burlington, Vermont; by steamer to Plattsburgh, New York; by train to Montreal, Canada; and then by steamer to Quebec City via the St. Lawrence River. As with his trips up the Merrimack, to Maine, and to Cape Cod, this trip by boat and rail was converted to travel writing in lecture and print. The result was *A Yankee in Canada*. By early 1851 Henry was back out on the lecture circuit again, this time with essays spun off from a second trip to Maine, "Walking" and "The Wild," which were later merged and published posthumously under the former title. Though these lectures were popular, they were never financially remunerative.

Thoreau's permanent commitment to a boating life was sealed when his family began settling into the Yellow House at 255 Main Street in September 1850. There he gained exclusive use of the third-floor garret running the length of the house, each side with its own window. Counter-intuitively, this sanctum was more private than his house in Walden Woods because it didn't attract visitors, didn't require daily trips into the village, and was fiercely protected by the women of the house. To keep it warm in the winter, he burned wood, much of which was driftwood that he collected from the river. His reflections in his journal of this aquatic task are reminiscent of his terrestrial foraging for stumps in the "House Warming" chapter of *Walden*.[29]

Along with driftwood, Thoreau burned coal in the Yellow House, a fuel then increasing rapidly in popularity. We know this from his account of coal deliveries by a teamster he nicknamed the "Littleton Giant" and his descriptions of the Carboniferous Period plant fossils the coal contained. "That fern leaf on my coal (?)," he remarked, "is probably the *Neuropteris* as figured in Richardson's Botany." As with the "mind-prints" of his Holocene archaeology (see Chapter 1), the ancient prints of this earthy substance helped him travel back in time.[30]

That winter Thoreau began reading much more extensively in natural history. In the spring he received a circular from the Smithsonian sent to scientists throughout the country. The circular, titled "Registry of Periodical Phenomena," asked "all persons who may have it in their power" to record their observations of "periodical phenomena of Animal and Vegetable life" and to "transmit them to the Institution." This marks the onset of Thoreau's commitment to all-season sojourning in order to inventory seasonal phenomena. It gave structure to what previously had been more open-ended explorations.[31]

Down the Drain

By early 1851 it was very clear that the Middlesex Canal would go bankrupt, and its directors wanted out. That October its corporation "voluntarily relinquished the use of, and closed, their said Canal, as a public accommodation."

And, for the trifling sum of $20,000, they sold to "Charles P. Talbot of Lowell, and Thomas Talbot, of Billerica . . . Manufacturers and Copartners . . . eleven pieces or parcels of land . . . the use of the waters of Concord River, and the mill privileges thereto." (Years later, Talbot would charge the government a quarter of his purchase price simply to allow the 1861 hydraulic experiments to be run for a few weeks.) One condition of the sale was to maintain high water behind the dam by never letting its surface fall more than three-quarters of an inch below the dam spillway. Using this closed-door legal maneuver, the Commonwealth let a private shareholder corporation convert the flow of the Concord River from a public good to a private source of wealth. By unanimous vote, they later requested of the Senate and House of Representatives that they "be discharged from the obligation to keep the canal open for navigable purposes," and to "surrender the franchise of the corporation." In legal-speak, they "voluntarily memorialized the Legislature to release the said Proprietors from any further obligation."[32]

This purely administrative dissolution of the canal rattled the meadowland farmers. Though the "Canal has been, for ten years, abandoned and broken up," the corporation's directors made a final profit by selling "their rights in this river, whether of feeding their Canal, driving mills, or boating to the Causeway." They conveyed "these water-privileges, but without warranty, to a person now in possession, who claims as a *mill*-privilege all that was granted as canal and boating privileges; consequently, the right of keeping up the Dam at its present height, and, of course, that of *increasing* it."[33]

This behind-closed-doors sale of Concord River water coincided with the sale of even more extensive water privileges in the Merrimack Valley. I refer to the slow takeover of New Hampshire's vast freshwater lakes by two major waterpower companies at Lowell and Lawrence, the PLC and the Essex Company, respectively. They combined to create the Lake Company, officially known as the Winnepissiogee Lake Cotton and Woolen Manufacturing Company, with each constituent company owning half the shares. Beginning in 1845, and slowly and covertly, they began to acquire the rights to New Hampshire's four largest lakes, Winnipesaukee, Squam, Wentworth, and Sunapee, so they could manage their summer flows for the mills of Lowell, and later their downstream counterparts in Lawrence. This action eventually led to a violent conflict over water rights during the summer of 1859 within a hundred miles of the peaceful but still simmering conflict in Concord.[34]

5 | Port Concord

IN JUNE 1851 HENRY STUMBLED onto a book that would change the course of his life: *Journal of Researches into the Natural History and Geology of the Countries Visited during the Voyage of H.M.S.* Beagle *Round the World, under the Command of Capt. Fitz Roy, R.N.* This was later published under the more parsimonious and marketable title *Voyage of the* Beagle. Its author was a young naturalist named Charles Robert Darwin, who for the first thirty years of his life was known mainly as a geologist. Thoreau's journal entry of June 7, 1851, is the first to mention Darwin's book. By June 11 he was transcribing copious extracts from it into his commonplace book, which gushed into a journal that would never be the same.[1]

Darwin's *Journal of Researches* would become Thoreau's favorite book within his favorite genre of travel writing. Finally he had discovered an author who also appreciated Alexander Humboldt, whose writing was compelling, who drew great insights from the minutest facts, and who stood in awe of what the "Author of Nature" had written. And for the first time Thoreau met through his readings a young scientific naturalist whose daily adventures often began with a short voyage in a skiff followed by the study and collection of natural history objects.

Within two weeks of having finished Darwin's report of his round-the-world ocean voyage, Thoreau went to the sea for an adventure of eight days. Arriving in Cohasset on July 26, 1851, he "called on Captain Snow," who regaled him with stories of great waves and told tales about his seafaring grandfather Jean (John) Thoreau. He had commanded "a packet between Boston or New York and England" and had also run a store on Long Wharf in Boston, where fishermen "'fitted out at Thoreau's.'"[2]

After walking south to Duxbury, Henry overnighted in a tavern, whose keeper was heading out to sea for a weeklong cruise to go "a-mackereling" in a forty-three-ton schooner with two sails and four dories. Spontaneously Henry joined the crew of seven, but only for part of the day. Following a

FIGURE 16. Two river voyageurs. Historic photograph by Alfred Munroe. Nashawtuc Hill, with meadows flooded, about 1892. Courtesy of the Concord Free Public Library.

visit to Clark's Island, he sailed on July 29 with his uncle Ned, experiencing a powerful storm: "A northeast wind with rain, but the sea is the wilder for it. I heard the surf roar . . . the rut of the sea." After sailing to Plymouth, they "landed where the Pilgrims did and passed over the Rock on Hedge's Wharf. . . . Saw many seals together on a flat." This trip permanently enhanced Thoreau's appreciation of sailing. He never forgot the thrilling roar of the sea, which he would later compare to the roar of local woods in a storm. "This sailing on salt water was something new to me. The boat is such a living creature, even this clumsy one sailing within five points of the wind. The sailboat is an admirable invention, by which you compel the wind to transport you even against itself. It is easier to guide than a horse. . . . I think the inventor must have been greatly surprised, as well as delighted, at the success of his experiment. It is so contrary to expectation, as if the elements were disposed to favor you."[3]

On July 30, he sailed once again with his uncle Bill, taking the helm of a large vessel for the first time in his life. It was a transforming experience.

En route to Duxbury, he wrote, "I was steersman and learned the meaning of some nautical phrases: to keep the boat 'close to the wind' till the sails begin to flap; 'to bear away,' meaning to put the sail more at right angles with the wind; a 'close haul,' when the sails are brought and belayed nearly or quite in a line with the vessel." These actions would later resonate in his journal. Coming back again via Plymouth Rock, he solidified his identity as a New Englander.[4]

After his return to Concord on August 1, Thoreau thereafter found the difference between land and sea diminished, especially during storms. "The wind roars amid the pines like the surf. . . . The whole country is a seashore, and the wind is the surf that breaks on it." In October he vivified this analogy when he "cut three white pine boughs . . . and set them up in the bow of our boat for a sail." For the rest of his life, he recognized that "all the land is a seacoast to the aerial ocean. It is the sound of the surf, the rut of an unseen ocean, billows of air breaking on the forest like water on itself. . . . The earth is our ship, and this is the sound of the wind in her rigging as we sail."[5]

ON A HOT AUGUST EVENING LATER THAT MONTH, and after a refreshing swim in the Sudbury River, Thoreau found himself in a particularly meditative state. While dripping dry beneath Bittern Cliff and "sitting on the old brown geologic rocks," he became a "continent of thought," taking in "the stillness and coolness that evening brings." In the distance, he heard "the grating of some distant boat, which a man is launching on the rocky bottom" to explore "the scenery of this river! What luxuriance of weed, what depth of mud along its sides! These old antehistoric, geologic, antediluvial rocks, which only primitive wading birds, still lingering among us, are worthy to tread," a world "like what we love to read of South American primitive forests." Here Thoreau was reflecting on Darwin's modus operandi of exploring his unknown shores by boat. During the next few years his daily practice would slowly evolve into something similar. Leaving the mother ship of his attic sanctum, he would launch a dory to explore Concord's three rivers and the lands beyond them.[6]

When the valley fogs of September arrived that year, Thoreau climbed above them to see this ocean of fog with its flat white top lying snug in the valley bottoms. "Oh, what a sail I could take, if I had the right kind of bark," he rhapsodized. "And all the farms and houses of Concord are at bottom of that sea." His land had become an ocean in yet another way.[7]

By late September Thoreau realized that the Concord River—still officially part of the defunct Middlesex Canal—was a busy inland port. At the town landing he saw "all kinds of boats chained to trees and stumps by the riverside,—some from Boston and the salt water." None, however, was "so suitable and convenient as the simple flat-bottomed and light boat that has long been made here by the farmers themselves," and which Thoreau had made three of in his youth: the *Rover,* the *Red Jacket,* and the *Musketaquid.* By mid-October he was classifying Concord's boats into "different patterns,—dories, punts, bread-troughs, flatirons, etc., etc.," none of which, not even those of the "Boston carpenters," could compare with the "prevailing . . . genuine dead-river boats." On the water he saw "water carriages of various patterns and in various conditions,—some for pleasure; some for ducking, small and portable; some for honest fishing, broad and leaky but not cranky; some with spearing fixtures; some stout and square-endish for hay boats; one canal-boat or mud-scow in the weeds, not worth getting down the stream, like some vast pike that could swallow all the rest, proper craft for our river."[8]

A few oversized canal boats still plied the waters of the Concord River: "Passed a large boat anchored off in the meadows not far from the boundary of Concord. It is quite a piece of ocean scenery, we saw it so long before reaching it and so long after; and it looked larger than reality." Also, the village barber had a "long boat which he built so elaborately himself, with two large sails." During a warm-water freshet on a Sunday in May, smooth water on the meadows brought "many out in boats. . . . All men and women who are not restrained by superstitious custom come abroad this morning by land or water, and such as have boats launch them and put forth in search of adventure." Though Concord was clearly no St. Louis or Pittsburgh or even Bangor, its waterfront had become an important part of its reality, especially for a boatman named Henry Thoreau.[9]

By late October 1851, Henry found himself dreaming of using his "own small pleasure-boat, learning to sail on the sea." Within a week, he discovered an inland sea large enough for his imagination, one that would never drain away through the Fordway: "Long Pond in Wayland, Framingham, and Natick, a great body of water with singularly sandy, shelving, caving, undermined banks. . . . It is a wild and stretching loch, where yachts might sail,—

Cochituate. It was not only larger but wilder and more novel than I had expected. . . . The shore suggests the seashore. . . . It must be the largest lake in Middlesex."[10]

This "wild and stretching" view of Long Pond was an engineered artifact of the nineteenth-century makeover. The City of Boston had raised the water level with a dam, creating a brand-new shoreline on a lake with a long fetch nearly parallel to the prevailing northwesterly winds, and with surrounding banks composed of soft glacial deltaic sediments. This had initiated a surge of shoreline erosion that was reaching its peak around the time he made this observation. Thoreau perceived the landslides and eroding banks as comparable to what he'd seen on Cape Cod, because the same geologic materials and forces were involved. Seeing these golden shoreline sands within his own watershed helped him bond to his native soil even more tightly.

THOREAU ALWAYS HAD ACCESS TO A BOAT OF SOME SORT because they were fairly common on the river. But it's hard to know how many he owned, rather than borrowed, and for how long. His sale of the *Musketaquid* to Nathaniel Hawthorne in August 1842 left an ownership hiatus not completely closed until ten years later, in August 1852. Within this hiatus, I've found no explicit reference to his owning a boat. During the last two years of this hiatus, William Ellery Channing, referred to as "C." in the journal, became his most frequent boating companion, if only because he had become an across-the-street neighbor. (This resident of Concord is not to be confused with his namesake from the previous generation, the founder of American Unitarianism.) Recall that Hawthorne had bequeathed *Musketaquid* to Channing in 1846, making it available for C. and Thoreau to share. This could help explain why Thoreau didn't use the phrase "my boat" in his journal prior to the fall of 1852.

During the apparent ownership hiatus Thoreau was in the habit of borrowing boats at various points on the river. When visiting Walden Pond, he also found one to use, though it's not clear whose, or where it was moored. Ditto for his river sojourns. On October 7, 1851, Henry and Channing borrowed one "made of the thinnest and lightest material, without seats or thole-pins, for portability." It was constructed from "three distinct boxes shaped like bread-troughs, excepting the bow piece, which was rounded, fastened together by screws and nuts, with stout round leather handles by which to carry the separate parts." In mid-October 1851 they borrowed one so "miserably built" that they "expended nearly as much time and labor in counteracting the boat's tendency to whirl round" as they did moving forward. They traveled "best and fastest when the wind is strong and dead ahead." The following season they "borrowed Brigham the wheelwright's boat at the Corner Bridge," of which Henry provided no description whatsoever. Once Henry and Channing used "an old leaky boat" that "leaked so,—faster and faster as it sank deeper and tipped with the water in it,—that we were obliged to turn to the shore." On another occasion they made use of a boat at Beaver Pond that leaked so badly they caulked it with newspaper. From these descriptions, I imagine a comedy of errors.[11]

By the following spring Thoreau and Channing were making frequent sojourns that began with a boat and ended on foot. "Channing calls our walks along the banks of the river, taking a boat for convenience at some distant point, *riparial* excursions. . . . We take a boat four or five miles out, then paddle up the stream as much further, meanwhile landing and making excursions inland or further along the banks." Such boat-assisted sojourns would become Thoreau's most common mode of exploration for the last eight years of his life. Thus, it was only a matter of time before he obtained a boat of his own.[12]

By August 31, 1852, Thoreau was writing about "my new boat." I've found no record of its construction, purchase, or description, so it was likely quite ordinary. Unlike the three boats he built in the 1830s, this boat (and possibly one more) may have been purchased or given to him. Perhaps it was that year's unusually high and long-lasting spring freshet that convinced him to get a boat of his own. We can safely assume it was like those he had built, a flat-bottomed dory built mainly for rowing in shallow water.

On that day he outfitted his new boat for some serious voyaging: "I rigged my mast by putting a post across the boat, and putting the mast through it and into a piece of post at the bottom, and lashing and bracing it, and so sailed most of the way." This describes a conversion from rowboat to sailboat. After giving it a test run, he began to describe himself as a sailor: "How much he knows of the wind, its strength and direction, whose steed it is,—the sailor . . . I find myself at home in new scenery. I carry more of myself with me. . . . If we sail, there is no exertion necessary. If we move in the opposite direction, we nevertheless make progress. And if we row, we sit to an agreeable exercise, akin to flying." Aesthetically, "a sail is, perhaps, the largest white object that can be admitted into the landscape." Henry's new boat had become an extension of himself.[13]

"It is pleasant to embark on a voyage, if only for a short river excursion, the boat to be your home for the day, especially if it is neat and dry. A sort of moving studio it becomes, you can carry so many things with you. It is almost as if you put oars out at your windows and moved your house along. A sailor, I see, easily becomes attached to his vessel. . . . We move now with a certain pomp and circumstance, with planetary dignity. The pleasure of sailing is akin to that which a planet feels. It seems a more complete adventure than a walk." By this stage, Thoreau had become so attached to his new boat that he considered it a second home, a floating sanctum beyond the Main Street house. He explicitly states that a sojourn by boat is a greater adventure than one with only boots on the ground.[14]

For the next four months (the freeze-up was late that year) Thoreau took his boat out on dozens of sojourns, learning to see the river scientifically in ways that he had not been able to before. By December 2 of that year he was describing himself as a "boatman" with enough experience to know the river by its different reaches. And at the very end of the year, on December 27, he took this new boat out to Walden Pond so that he could paddle across the water, bringing it back the next day in the rain. The best explanation I can come up with for this round-trip trouble is that he felt the compulsion to boat even after the river was frozen. After that trip, he put his boat into storage at the family home, almost certainly in the cellar.[15]

After he got his new boat in August 1852, Thoreau's journal entries about borrowing boats disappear, except for rare later exceptions, such as the time he had to borrow one to retrieve his boat after it floated away. Notably, Thoreau's explicit mentions in the journal of sojourning by boat rise dramatically, from fifteen in 1852 to a lifetime peak of sixty-four in 1853. For his last seven complete years of good health, Thoreau boated Concord's waters at least fifty-five times each year. This is five times as often as the average American boat owner goes out on the water today.

When he relaunched his new boat the following spring, on March 8, 1853, it leaked, perhaps as a result of the bouncing journey to Walden Pond at the close of the previous season. This forced a repair on March 14. By March 19 he had repainted it with paint he mixed himself, putting the dry pigment through an old coffee grinder and mixing it with oil: "Spanish brown and raw oil were the ingredients." Spanish brown was the "ubiquitous color" of early

America, "dirt cheap because it is dirt," being composed of iron oxides from the soil. Thoreau may have dug up this pigment himself and then ground it to the texture of fine espresso. In that era, paint was sold in components to be mixed before use: generally the oil, the turpentine, and whatever pigment one had on hand.[16]

When the paint dried, Thoreau "filled the seams with some grafting-wax I had melted." He then "broke up the coffee-mill" and used its metal to protect the prow of his flat-bottomed boat. On March 22 he announced: "Launched my new boat." (His ownership was still less than one year old.) "It is very steady, too steady for me; does not toss enough and communicate the motion of the waves. Besides, the seats are not well arranged; when there are two in it, it requires a heavy stone in the stern to trim. But it holds its course very well with a side wind from being so flat from stern to stern." The report continues: "My boat is very good to float and go before the wind, but it has not run enough to it,—if that is the phrase,—but lugs too much dead water astern. However, it is all the steadier for it. Methinks it will not be a bad sailer." Sailing had become so much a part of his daily routine that he felt compelled to take a "cruise" before heading off to work for a day of surveying on March 31, 1853. He was out on the water and back before 9:00 AM.[17]

Around the same time, Thoreau found a suitable "harbor" in the "willows" on Ellery Channing's land. This was his first explicit mention of the place, hereafter known as the "boat place," located on his scroll map as the western, upriver end of the boat-place bar. Though he kept his boat partially concealed, thefts became an issue. On July 21 he looked for the "rascals" that had stolen his seat. Years later he spent an entire day looking for the whole boat, which had been taken. During his riparial excursions, he would hide his boat in riverside pullouts, and cache his sails and oars where they would not be pilfered while he went on the terrestrial parts of these sojourns.[18]

Sailing under different conditions required daily adjustments and maintenance. He needed stones for ballast, and rollers to move his boat up and down over causeways should the bridges be flooded. "Paddling" and "rowing" became synonyms. Sails were used whenever possible. One day he and Channing sailed under the power of one sail and two umbrellas. After breaking his mast because he'd misjudged the height of a bridge, Thoreau paddled up the Assabet to cut a new one.[19]

That spring Henry claimed ownership of the river: "Without being the owner of any land, I find that I have a civil right in the river,—that, if I am not a landowner I am a water-owner. It is fitting, therefore, that I should have a boat, a cart, for this is my farm." Explicitly, his river stands in for the farm he would never own, and his boat stands in for the cart that every farmer needs. "Since it is almost wholly given up to a few of us, while the other highways are much travelled, no wonder that I improve it. Such a one as I will choose to dwell in a township where there are most ponds and rivers and our range is widest." Explicitly, he's choosing to stay in town because of its rivers.[20]

"In relation to the river, I find my natural rights least infringed on. It is an extensive 'common' still left. Certain savage liberties still prevail in the oldest and most civilized countries." These sentiments, so often assumed to refer to the rocky Easterbrook country, Beck Stow's Swamp, and Walden Woods, instead refer to Concord's three rivers. Ironically, these riparian liberties were then in the process of being preserved by the railroad, which had shifted commerce away from the water toward overland transportation.[21]

Each of Thoreau's three rivers became a chain of self-named personal destinations for his sojourns. In a reminiscence from 1855, he listed these on the Sudbury River: "The Rock (at mouth), Berrick's Pasture, Lee's Hill, Bridge, Hubbard Shore, Clamshell Hill and fishing-place, Nut Meadow Brook, Hollowell Place and Bridge, Fair Haven Hill and cliffs, Conantum opposite, Fair Haven Pond and Cliff and Baker Farm, Pole Brook, Lee's and Bridge, Farrar's or Otter Swamp, Bound Rock, Rice's Hill and 's [*sic*] Isle, the Pantry, Ware hill, Sherman's Bridge and Round Hill, Great Sudbury Meadow and Tall's Isle, Causeway Bridges, Laurel Brook, the Chestnut House, Pelham Pond, the Rapids." The Assabet and Concord Rivers have comparable lists, which I don't mention here for brevity. His sojourning country included a collection of nearly a hundred waterside places.[22]

During the rest of Thoreau's life, his respect for the river system as a wild commons strengthened: "I think that I speak impartially when I say that I have never met with a stream so suitable for boating and botanizing as the Concord, and fortunately nobody knows it. I know of reaches which a single country seat would spoil beyond remedy, but there has not been any important change here since I can remember." Only as a result of his river study in 1859 did Henry get a chance to accurately measure how big this commons was. "This 16 miles up [river] added to 11 down [river] makes about 27 that I have boated on this River—to which may be added 5 or 6 miles of the Assabet—The sum is at least thirty-two continuous miles of river water experiencing constant change.[23]

Henry's global nautical interests suffuse the writings he is best known for. When living at Walden Pond, seven years earlier, he was "refreshed and expanded" by the smells of far-off lands drifting down from the railroad: "I smell the stores which have been dispensing their odors . . . which remind me of foreign parts of coral reefs & Indian oceans and tropical climes—& the extent of the globe—I feel more a citizen of the world at the sight of the palm leaf."[24] When *Walden* was in press, he visited Long Wharf in Boston. "As I watched the various craft successively unfurling their sails and getting to sea," he sighed, "I felt more than for many years inclined to let the wind blow me also to other climes." Though this never happened, it's easy to imagine him signing up as a seaman, as did two of his literary contemporaries, Herman Melville, author of *Moby-Dick*, and Richard Henry Dana, author of *Two Years before the Mast*. Instead, Thoreau made a commitment to sail the transient freshwater seas of Concord in flood.[25]

Modes of Boating

West Main Street in Concord was typical of its day. A hard-packed dirt road on which horse-drawn carriages, farm wagons, and pedestrians traveled east and west, it was redolent with manure all year, frozen in winter, muddy in spring, dusty in summer, and littered with leaves in the fall. Crossing that road twice a year was Henry's boat, making the trip between household winter storage and his own miniature version of Port Concord. The means of conveyance is not known but was probably some combination of a specially built wheeled cart and portaging. Once he had gotten the boat over the road, the path headed alongside Channing's house, through that property's back garden, and then down to cold water at the willows.

Like the day boaters of today who trailer their boats to a summer moorage, Thoreau got his boat to the water as soon as weather would allow, usually in early March. He put it back into storage as late as possible, usually in mid-December. Early April might find him poling his way through softened ice floes, following leads blown open by the wind. Early December might find him chopping ice out of his boat to find a place to sit. He appreciated the off-season because it intensified the pleasure of getting back on liquid water and gave him a chance to repair his boating gear. "I love to have the river closed up for a season and a pause put to my boating, to be obliged to get my boat in. I shall launch it again in the spring with so much pleasure. There is an advantage in point of abstinence and moderation compared with the seaside boating, where the boat ever lies on the shore. I love best to have each thing in its season only, and enjoy doing without it at all other times."[26]

He imagined "the different styles of boats that have been used on this river from the first, beginning with the bark canoe and the dug-out, or log canoe, or pirogue." These were the boats of the Holocene. The typical boat of the Anthropocene "probably has its prototype on English rivers,—call it dory, skiff, or what-not,—made as soon as boards were sawed here; the smaller, punt-like ones for one man; the round-bottomed boats from below, and the half-round or lapstreaked, sometimes with sails; the great canal-boats; and the hay-boats of the Sudbury meadows; and lastly what the boys call 'shell-boats,' introduced last year, in imitation of the Esquimaux kayak."[27]

On at least a hundred occasions Henry uses the pronoun "we" to describe a boating sojourn without naming his companion. We can presume it was Channing. His sister Sophia was next in frequency, with Bronson Alcott, William Tappan, Theophilus Brown, William Blake, R. W. Emerson, E. Hoar, and Daniel Rickertson joining him more than once. All extant members of his family joined him, listed as Father, Cynthia, Sophia, and Aunt. By count forty-five others joined him on the water, including Channing's "pup," several unnamed ladies he took "a-graping," and an anonymous Irishman. Henry often went alone, even when others asked if they could come. And to the extent possible, he avoided being seen by his fellow townsmen, which sent him up the Assabet more often than not.[28]

AFTER YEARS OF EXPERIENCE, Thoreau had become an expert local meteorologist. The prevailing winds, as indicated by the weather vanes on hilltop farms, were usually northwest and southwest. These winds were confined and strengthened within the valley, especially in narrows, meaning that Thoreau could "always sail either up or down the river with the rudest craft, for the wind always blows more or less with the river valley." This was due to the concentration of the airstream and the reduction in wind friction over water, relative to the riparian woods. The result was a river of air tracking the river of water below. For example, he discovered he was able to sail through all the "windings" of the upper Sudbury River with a steady breeze, even though the meanders were tightly looped. In spring and fall, the river generated its own cross-breezes and gusty winds, being colder or warmer than the adjacent land. With light winds and a confined reach, he found that the "wind and current are almost evenly matched." In this situation, he could "sit carelessly waiting for the struggle between wind and current to decide itself," which would determine whether he would go upstream or down. This is a close parallel to Thoreau's experience with local winds at Walden Pond, where he let the radial zephyrs decide which shore to send him to.[29]

Though the prevailing winds were very predictable on the open meadows, they were "much affected by" the hills, woods, and zigs and zags within the valley, especially on the Sudbury. These circumstances created reaches where the wind seemed to lull completely, or to blow counter to expectations, which Thoreau found "baffling." Generally, this was up-valley for a northwest wind, and down-valley for a southwest wind. These short reaches lay on the blunt lee sides of hills projecting in the direction of the wind. Thus to zig and zag up or down the river often required some combination of sailing and rowing.[30]

Thoreau typically sailed sitting down. In fact, one distinct advantage of sailing over rowing was that he could write in his notebook while traveling (a practice far less dangerous than texting while driving). Under windy conditions, or when he wanted to make good time, he used the racing techniques of a yachtsman: "I sailed swiftly, standing up and tipping my boat to make a keel of its side," and "I must keep it pointed directly into the teeth of the wind. If it turns a little, the wind gets the advantage of me and I lose ground." He loved sailing in the strongest winds, those that required "all my strength and skill" to "turn about."[31]

Sailing in a complex wind regime through three seasons every year brought him many challenging experiences. At times he became as miserable as any sailor can be without giving up. Occasionally he would get stuck on "low spits of grass ground," meaning meadow, meaning some combination of mud, sand, and wetland vegetation, and he would have to get out and pull himself back into deeper water. In one early April storm, high, frothy whitecaps drove water sideways into his boat, heavy rain poured down from above, and the boat leaked from below. This forced Channing and him ashore every "half-hour" to "upset" their boat, tipping out the water before carrying on. After that, the rain turned into heavy wet snow that made them "white as millers and wet us through."[32]

On another occasion in late April, Henry was alone. A very heavy rain squall forced him to turn about in Fair Haven and head home. With unfavorable winds, he raised his sail and cowered in the stern beneath an umbrella he held with his knees while simultaneously steering with the rudder and paddling. With the strong easterly wind that afternoon, he was repeatedly driven to the western shore. Giving up, he lowered his sail and rowed with his back to the horizontal rain all the way home. During a bitterly wet and cold November storm, Channing and Thoreau found themselves crosswise to the wind, which blew water over the gunnels and repeatedly pushed them into the "button-bushes" on the shore. They had to row out into the teeth of the wind, right the sail, and sail until blown back again. Fully soaked and "indifferent to wet," they "took to rowing vigorously" to keep warm, reaching home "just after candlelight."[33]

On one occasion Thoreau nearly capsized his boat. As he wrote: "It is a strong but fitful northwest wind. . . . Under my new sail, the boat dashes off like a horse with the bits in its teeth. Coming into the main stream below the island, a sudden flaw strikes me, and in my efforts to keep the channel I run one side under, and so am compelled to beach my boat there and bail it."[34]

The biggest single mistake of his boating career that he confessed in writing was having his boat drift away in the wind. Stepping ashore to examine a dead sand turtle one day, he turned around to find that his boat was "half way across the river, blown by the S. W. wind." His solution was to walk upstream, borrow a boat, row down to catch his, tether his to the bank, row back upstream, return the borrowed boat, and then walk back down to his own. This lesson helped him realize that, without a boat, the river was a remarkable "bar" that made him "comparatively helpless." Thus it was that the boatman and his boat became inseparable.[35]

PADDLING (WHICH TODAY WE CALL "ROWING") was his next best alternative to sailing, and clearly his default mode of travel. His oars he called paddles, and his oarlocks he called thole pins. Some days were all rowing, others were all sailing, but most required some combination of the two. In the spring he appreciated the "exercise of the arms and chest" rowing gave him "after a long winter's stagnation during which only the legs have labored." In summer, however, that exercise caused him to overheat. Thoreau learned a trick for cooling down on river voyages when he didn't want the bother of landing, undressing, swimming, dressing, and reembarking: when passing beneath bridges, he would hang on to the abutments for a while, and the combination of shade and a steady current of cool air circulating over the water would cool him down nicely. There were twenty such cooling stations on Musketaquid alone.[36]

Some of his excursions were of remarkable length. From his boat place he would sometimes row upriver to Sudbury and back, "twenty-five miles to-day." Clocking his average speed, he found it to be over three and a half miles per hour, equivalent to a brisk walk. The main problem with rowing, especially in wide reaches, was that strong crosswinds would force his boat to "whirl round." This he countered by zigzagging such reaches, comparable to tacking when sailing.[37]

Paddling didn't work when the water was too shallow, because the oars didn't get enough depth to pull, or when the channel was too narrow to extend the oars outward. At such times Thoreau used a pole to push himself over shallow water into hidden alcoves. When the meadows were in flood, the water depth for nearly a mile might be no more than a few inches, making poling a far better mode of propulsion than rowing or sailing. This was especially true when the pattern of sedge tussocks and tufts dropped in the meadow by ice created a labyrinth. Poling by pushing was also the best way to get through open leads between ice floes.

On the main channel in midsummer, the stage often dropped to the point where the slow, stagnant river shoaled over sand bars that had been enlarged and raised by sediment pollution. Making this worse was the dissolved nutrient pollution associated with land clearing, the decay of soil humus in pastures, and livestock feces. This stimulated the growth of aquatic plants to the point where the green scum of algae and duckweed grew thickly. Above shallow sediment bars, larger weeds choked the river to such an extent that Henry claimed his oars caught a "crab," or got stuck on every downward stroke and became draped and trailed by weeds on the way up. He reported weeds so thick that frogs and web-footed birds could hop or walk across the full width of the river without getting wet. In those situations, stepping overboard and pulling worked far better than poling.[38]

And then there were weedy patches described as so thick that each acted like a dam with millions of holes in it, something like a colander, sieve, cheesecloth, or filter paper. Water backup behind them was proven during the 1861 experiments, when hacking away the weeds speeded up the current and lowered the water a few inches in upstream meadows. Such weeds could not be rowed through, or even poled through. Instead Thoreau jumped in, got wet up to his knees, and dragged his boat to deeper water. On hot summer days he compared this scenario to the "river of wailing" in Greek mythology, the Cocytus, portrayed in Dante's *Divine Comedy* as one of many pathways to hell. One drought was so severe that Thoreau spent half an hour hauling his boat over clamshells growing in situ on the bottom of the Sudbury River. If true, this indicates only a trickle of water.[39]

During late winter breakup, Thoreau found himself dragging his boat over ice jams and floes to reach open water. During freeze-up, he would sometimes have to chop his boat loose from the thin ice and then skid it out toward the middle to reach navigable water. During late-summer droughts, falling river stage would expose rocks and snags that had been unseen for years. If stuck, he would have no choice but to drag himself off.

Bridges were a chronic problem because high water meant low arches. "At midsummer I can sail under them without lowering my mast," he wrote, but at times of high water the arches of bridges "were completely concealed by the flood," obliging him to "find a round stick and roll my boat over the road." Under moderate conditions he simply lowered his mast and floated beneath. Sometimes it was a tight squeeze.[40]

THOUGH BOATING WAS THOREAU'S MAIN MEANS OF RIVER TRANSPORT, he also accessed the river by skating and walking. Once he discovered that the snow was so ice-crusted that he could skate up and down the landscape far away from the river. Even better conditions were reported by Rufus Hosmer for 1820, when the snow was so deep and the glaze so thick that "you could skate everywhere over the fields and for the most part over the fences."[41]

The "rapidity" with which Henry could skate on the river astonished him: "In a quarter of an hour I was three and a half miles from home without having made any particular exertion,—*à la volaille*." Taking a twelve-mile-long skate was nothing out of the ordinary. On January 31, 1854, conditions were so perfect that he skated a total of sixty miles up and down! Because each bridge usually had a stretch of open water, he often had to use "bits of wood with a groove in them for crossing the causeways and gravelly places that I need not scratch my skate-irons."[42]

Skating with the wind was an absolute delight. "I skated so swiftly before the wind, that I thought it was calm, for I kept pace with it, but when I turned about I found that quite a gale was blowing." One time Henry and a companion came downriver "with the wind and snow dust, spreading our coat-tails, like birds. . . . I found that I could sail on a tack pretty well, trimming with my skirts. Sometimes we had to jump suddenly over some obstacle which the snow had concealed before, to save our necks." On another occasion Henry sail-skated by resting the mast of his boat on his hip, holding the middle, and running before the wind on the meadow in front of his house, though he "could not easily tack."[43]

Skating was easily compromised by less than ideal conditions. Snow halted it entirely. A melt made it dangerous, even if the water drained off, because it created "shallow puddles on the ice . . . not easy to be distinguished," that "detained your feet while your unimpeded body fell forward." Alternatively, refreezing preceding drainage could create a "thin crispy ice on top of the old ice" that it was impossible to glide through. Worst was an unrecognized crack. Once, upon hitting one, he "slid on my side twenty-five feet."[44]

6 | Wild Waters

WALDEN WAS PUBLISHED ON AUGUST 9, 1854. After nine years of work on seven drafts written in two great spurts of energy, Thoreau's postpartum letdown was inevitable. To help him get through the following winter, he read *New England's Prospect* by the early colonial adventurer William Wood. In it he found fascinating descriptions of the Concord landscape prior to European settlement. In an oft-cited journal entry for January 24, 1855, he compared Wood's observations from the 1630s with his own from the 1850s. During initial Puritan settlement two centuries earlier, Musketaquid's river meadow grasses had grown more "rankly," meaning more luxuriantly. In Thoreau's era, previously abundant berries had been "cornered up by cultivation." The magnificent old-growth forests were gone. And his before-and-after lists of quadrupeds—birds, fish, reptiles, amphibians, and mammals—showed massive local extinctions. The losses were dramatic.[1]

Fourteen months later, on March 23, 1856, he revisited that list with a lament: "But when I consider that the nobler animals have been exterminated here,—the cougar, panther, lynx, wolverine, wolf, bear, moose, deer, the beaver, the turkey, etc., etc., I cannot but feel as if I lived in a tamed, and, as it were, emasculated country. . . . Is it not a maimed and imperfect nature that I am conversant with? I am reminded that this my life in nature, this particular round of natural phenomena which I call a year, is lamentably incomplete."

Later in this passage, his lament deepens beneath ecology to a sweeping generalization about the "whole" of the landscape, including its hydrology. "The whole civilized country is to some extent turned into a city, and I am that citizen whom I pity. . . . Primitive Nature is the most interesting to me. . . . All the great trees and beasts, fishes and fowl are gone. The streams, perchance, are somewhat shrunk."[2]

What began as winter reading became then-and-now comparisons, which revealed losses, which led to grief, which descended into pity, which

Thoreau turned inward on himself. This psychological downward spiral was featured by the historian Donald Worster in his magnificent 1977 book *Nature's Economy*. Six years later William Cronon used the episode to poignantly open his 1983 book *Changes in the Land,* a landmark in environmental history. The title of his opening chapter, "The View from Walden," works well as a retrospective literary device. But as plain fact, it seriously distorts the place, time, and context of Thoreau's message.[3]

Walden Pond, the alleged place of the lament, is an isolated, stone-rimmed kettle sunk deeply into an elevated gravel terrace. Thoreau's temperature measurements from August 23, 1860, would later prove that it was unique among local ponds: a "deep green well," a "genuine & cold spring" with unusually clear water and an unusually ascetic flora and fauna. Pond scum (the title of a recent critique of Thoreau's legacy) cannot happen because the water holds too little in the way of nutrients. Despite its woods being axed for fuel and its shore being nicked by the railroad tracks, Thoreau declared Walden immune from human transmogrification at the time scale of his own life: "It is itself unchanged. . . . It has not acquired one permanent wrinkle. . . . It is perennially young."[4]

Deforesting Walden was thus like getting a haircut. Though visually arresting, the trees grew back quickly, leaving a comparable scene within a few decades, and only a little more silt in the bottom of the pond. Indeed, the stability and eternal purity of Walden, when compared with the rest of his local geography, is a central theme of Thoreau's literary masterpiece. Leo Marx concluded that *Walden* the book effectively "removes" Walden the pond "from history." Today the pond shore is fully wooded and its water remarkably clean. Prior to the massive engineering works of the twentieth century, it was not unlike what Henry's "youthful eyes fell on" when he was a five-year-old boy seeing it for the first time. In fact, the "arrowy pines" he cut for the timbers of his one-room house were chopped from recent second growth. Walden Pond has been amazingly resilient to long-term changes in the land, making it an anti-icon for the Anthropocene.[5]

The actual place for Thoreau's lament was the third-floor attic of the family home on the main street of a bustling market town. The actual time of the lament was nine years after Thoreau left the pond to live at Emerson's house. By then he had lived in three subsequent domiciles, and the stone foundation of his one-room house in the woods had become archaeology. The actual context for Thoreau's lament was the "whole civilized country" of the Concord River Valley: a watershed draining 377 square miles and thirty-six towns in the process of being industrialized. Whereas the unprofitable Walden Pond had remained an outback in the woods owing to its dry, sterile soils and elevated topography, the profitable alluvial valley of the Concord River had been put to so many good uses that it was experiencing a nearly complete makeover.[6]

Thoreau's interest in "primitive nature" grew as it disappeared around him. And he knew what it looked like and felt like up close, having seen the emptiness of the deep Maine woods, the titanic sublimity of Katahdin, the deadly surf at Cohasset, and lightning bolts powerful enough to explode great trees up from their roots. Beginning in the late 1840s, he made a deliberate choice not to live the life of a pioneer or to join an Indian tribe. Instead he chose to be a "ruralist," as Lawrence Buell described him, a "Harvard-educated and genteelly subsidized" devotee of nature whose sister Sophia was his "most intimate collaborator in natural history pursuits." To quote Laura Dassow Walls, Thoreau "situated himself on the outskirts of town, as the active mediator of civil and wild." He preferred what Richard Judd calls the "second-nature" landscape of settled New England, rather than the "first nature" of untransformed regions with sparse human settlement. He chose to live in a family home run by his mother, with a meal on the table, a fire in the hearth, a piano in the parlor, and a kitten on his lap. When Henry lamented that his whole world had been "to some extent turned into a city," he was writing from Main Street, USA. In fact, his best friend and neighbor, William Ellery Channing, teased him that their Main Street houses opposite each other were little different from those found in the main streets of cities. Throughout his life, Henry sought the wild within a tamed landscape, rather than the tamed within a wild one.[7]

Thoreau's interest in primitive nature had always been mainly historical, as demonstrated by his lifelong obsession with archaeology, Native American ethology, and the seventeenth-century pioneers of colonial settlement. It was also spiritual, part of his quest to find his proper place in nature rather than in the Christian church he had been baptized in and would be eulogized in. And it was sensual, given his bodily needs to be outdoors on daily excursions. On this last point, he confessed: "I keep out of doors for the sake of the mineral, vegetable, and animal in me."[8]

Portrait Painting

In his twenties Thoreau had willingly given up his teaching posts. For more than a decade he earned his income mainly as a day laborer: digging ditches, laying bricks, building fences, building walls, doing carpentry, gardening, painting houses, and, in one celebrated case, shoveling manure. "On the tax lists," wrote historian Robert Gross, he was listed "as merely one among many landless laborers." Beginning about 1850, and driven by the urgency of paying off his debt for *A Week,* he chose land surveying as the day job best suited to his needs. He did so knowing it was considered a branch of civil engineering and that it was a critical step in the ownership, commodification, and development of land. Andro Linklater, in his historic analysis of surveying history, *Measuring America,* likened land surveying to a "wand" that must be passed over raw land before it can be "improved." Historian Daegan Miller called surveyors "the advanced guard of modern capitalism." Thoreau was a leader in that vanguard waving the magic wand of his surveying pole and sighting it with his compass.[9]

Before a factory mill can be built, the property must be legally bounded, and the surface topography must be characterized before engineering design can begin. Henry did this for at least four properties, thereby contributing to the industrialization of his rivers. Two of these surveys, those for the "Edward Damon factory site" and a "factory village in SW Concord," were on his favorite river, and interrupted his 1859 river project. Thoreau had long been keenly interested in the mechanical engineering of mill factories, especially the calculations of mill power based on the product of river discharge and its vertical fall. In fact, it was Thoreau's interest and aptitude in mechanical engineering that made the family's pencil and graphite manufacturing business profitable.[10]

Before a woodlot can be sold, its acreage must be measured so that its commodity value as a fuel can be accurately estimated. He did this dozens of times, especially for his townsmen thereby contributing to local deforestation. Before a farm can be subdivided for housing, a survey was legally required. Before an upland swamp can be redeemed for tillage, it must be drained. And with large drainage projects, accurate surveys were needed to determine the best pathways and gradients for flow. Thoreau helped kill several of the swamps he otherwise claimed to cherish.

In short, Thoreau personally and significantly contributed to the intensification of private capital development throughout the valley. Additionally, he surveyed for roads, cemeteries, and public buildings, which required the cutting away of hills and the filling of wetlands. Like the bankers, lawyers, builders, farmers, and elected officials who were his clients, Thoreau was an instrument of change. He knew it, and it made him uncomfortable. But he kept doing it anyway, because he needed the money.

His first paid survey, in November 1849, was to measure a woodlot for Isaac Watts. This job—performed with a compass borrowed from Cyrus Hubbard—followed the May 1849 publication of *A Week* by only five months. This was just enough time for its commercial failure to sink in, and for Thoreau to plan how he would pay down the debt. His decision in 1850 was to print an advertising broadside to launch his new business as a surveyor. Within one year he had plenty of clients and was seeing himself as an incarnation of Satan standing over his surveying stones. Within two years he was complaining to himself that his clients neither understood nor appreciated his true talents.[11]

"I am frequently invited to survey farms in a rude manner, a very insignificant labor," he wrote in 1852, "but I am never invited by the community to do anything quite worth the while to do." As a consequence, he performed his surveying with little enthusiasm and frequent complaint in his private journal. Though he clearly enjoyed the mathematical precision and the outdoor settings that came with this vocational choice, he considered client-driven land surveying to be his "portrait-painting"—his "art" that wasn't really art. As with the majority of service jobs in any economy, Thoreau did this to cover his room and board, the costs of travel, and incidental expenses. His true career goal was to earn the bulk of his income from the fine art of writing commercially successful books and giving public lectures to ticketed audiences. This never happened, so he kept on with surveying, somewhat begrudgingly, to near the end of his life.[12]

Thoreau admitted his complicity—if not hypocrisy—in a journal entry for November 11, 1850. It takes the form of a biblical allegory in which he is the "Prince of Darkness" measuring the "bounds" of the "worldly miser" to make "improvements" to Nature. In "Walking," Thoreau's much later rewrite of this passage, Satan the surveyor helps to "deform the landscape, and make it more and more tame and cheap." It's no wonder that Thoreau's journal occasionally overflows with invective after spending too much time with his philistine worldly employers. Thoreau didn't publish this allegory during his lifetime; instead, he worked with his sister Sophia to publish it posthumously in his most famous nature essay, which is mainly about the wild. This intentional delay makes it a deathbed act of contrition for his role as a land developer.[13]

NOT UNTIL AUGUST 30, 1856, did Thoreau resolve the tension between his yearning for primitive nature and his role in helping to destroy it. When searching for native cranberries on a cool early autumn day, he found himself in a patch of ground in the far corner of Beck Stow's Swamp that was so worthless that it had been left untouched by human hands, or so it seemed. "I see that all is not garden and cultivated field and crops, that there are square rods in Middlesex County as purely primitive and wild as they were a thousand years ago . . . little oases of wildness in the desert of our civilization. Is Nature so easily tamed? Is she not as primitive and vigorous here as anywhere?" The "here" being referred to was an "acre of secluded, unfrequented, useless (?) quaking bog," a place left untouched because it lacked all economic utility. "I could be in Rupert's Land," Thoreau wrote, comparing that tiny patch to subarctic Labrador, "and supping at home within the hour! This beat the railroad."[14]

Symbolically, the wildness of the red cranberry beat the civility of the steely locomotive. Organic, spontaneous, nonlinear, self-organized nature beat the metallic, engineered, linear, human-organized railroad. If we take Thoreau at his word, this insight reads like a religious conversion comparable to what struck him near the summit of Mount Katahdin: "I felt a shock, a thrill, an agreeable surprise in one instant, for, no doubt, all the possible inferences were at once drawn, with a rush, in my mind." After this integrative "rush," Thoreau's lamentation was gone for good.[15]

After rethinking his worldview, Thoreau dove straight to the heart of the matter: "It is in vain to dream of a wildness distant from ourselves. There is none such. It is the bog in our brain and bowels, the primitive vigor of Nature in us, that inspires that dream. I shall never find in the wilds of Labrador any greater wildness than in some recess in Concord, *i.e.* than I import into it. A little more manhood or virtue will make the surface of the globe anywhere thrillingly novel and wild. That alone will provide and pay the fiddler; it will convert the district road into an untrodden cranberry bog, for it restores all things to their original primitive flourishing and promising state."[16]

His explanation is crystal clear. The wildness of nature lies within our minds, not in our landscapes. Using his "brain," Thoreau could restore "all things" to their original state, even the "district road," by seeking and appreciating the wildness lurking beneath the appearances of the landscape—one he was helping to citify as a land surveyor, and as a business partner in Thoreau and Son, which had expanded from producing black lead merely for pencils to providing this form of graphite for the electrotyping industry. By deliberately choosing the proper paths along the river, he could "enjoy the "retirement and solitude of an early settler . . . and yet there may be a lyceum in the evening, and there is a book-shop and library in the village, and five times a day I can be whirled to Boston within an hour." So instead of focusing on his role in landscape degradation, Thoreau paid the "fiddler" by focusing on the novelty bubbling up everywhere, even the "district road," where wild-flowers sprang up in special abundance. In the case of the Assabet's pearly skiffs, he shifted his attention away from the miller's wastewater to the delight of the muskrat's midnight meals.[17]

Three years later, in the lead-up to his role in the flowage controversy in 1859, Thoreau realized that his own particular genius was also a manifestation of wildness. For an analogy, he turned to the chicken-thieving hen-hawk. "That bird will not be poultry of yours," he told his townsmen in his mind. "Though willed, or *wild,* it is not willful in its wildness." True wildness is never intentional. It simply is. It is an inherent property of all nature, including human nature. "It has its own way and is beautiful" in and of itself. In the human context, wildness is the true source of "any surpassing work of art," the impulsiveness of genius itself. "No hawk that soars and steals our poultry is wilder than genius," he concludes. Indeed, it was Thoreau's wild genius that most endeared him to his contemporary Walt Whitman, who referred to "his lawlessness—his dissent—his going his own absolute road let hell blaze all it chooses." This was consistent with Thoreau's own views on literature: "In literature it is only the wild that attracts us. Dullness is only another name for tameness."[18]

"In Wildness is the preservation of the World." This is arguably Thoreau's most famous quote, at least for bumper stickers. In the context of his hen-hawk epiphany above, it could be rendered as "In genius is the preservation

FIGURE 17.   Choppy sea. Historic photograph by Herbert Wendell Gleason.
At Red Bridge, Concord, Massachusetts, March 29, 1920. Courtesy of the Concord Free
Public Library.

of the world." By "genius," he meant his own instinctive, intuitive, individu-
alistic, impulsive being. By "world," he was referring to the one we create
with our minds. This was transcendentalism in its purest sense. In the process, Thoreau could easily convert his flooded meadows into the wild waters
of an inland sea.[19]

Inland Sea

Long before Thoreau's lifetime, the alluvial plain of the Concord Valley
lay at the bottom of a gray glacial lake. This beaded ribbon of turbid water
extended the whole length of the valley, widening over bedrock basins that
would later become meadows, and narrowing in bedrock constrictions. In

Thoreau's epoch, every strong flood recreated the moccasin footprints of this ancient glacial lake at a lower level. The result was a "chain of handsome lakes" that was made higher, more frequent, and more long-lasting by the direct and indirect effects of the Billerica dam. He described the largest lake, over the Sudbury Meadows, as a "smaller Lake Huron," more than a mile across in every direction. Next in size was that over the Great Meadows of Concord, more than two miles long and half a mile wide. Both of these transient lakes could last for weeks at a time, which was long enough for him to be surprised when they finally disappeared. During floods, the already wide Carlisle reach expanded to resemble one of New York's smaller Finger Lakes.[20]

"The element of water is in the ascendant," wrote Henry excitedly during a typical spring freshet. "Many new islands are made,—of grassy and sometimes rocky knolls, and clumps of trees and bushes where there is no dry land." During these times, "there is absolutely no passing, in carriages or otherwise. . . . Of eight carriage roads leading into Concord, the water to my knowledge is now over six." With such a vast expansion of gray, stormy water, the wildness of Nature put the more feeble power of humans into perspective.[21]

Thoreau often imagined Concord's inland ocean complete with many kinds of seabirds. "They come annually a-fishing here like royal hunters, to remind us of the sea, and that our town, after all, lies but further up a creek of the universal sea, above the head of the tide. So ready is a deluge to overwhelm our lands, as the gulls to circle hither in the spring freshets. To see a gull beating high over our meadowy flood in chill and windy March is akin to seeing a mackerel schooner on the coast. It is the nearest approach to sailing vessels in our scenery. . . . Oh, how it salts our fresh, our sweet-watered Fair Haven all at once to see this sharp-beaked, greedy sea-bird beating over it. For a while the water is brackish to my eyes. It is merely some herring pond, and if I climb the eastern bank I expect to see the Atlantic there, covered with countless sails."[22]

With high water and high winds came high times for Henry the navigator. He seized every opportunity to go sailing, not only during the spring freshet but also during the late summer hurricanes in August and September and the extratropical cyclones of winter nor'easters. For such large bodies of water, the mood was one of "angry dark waves," muddy whitecaps, bubbly streaks of dirty foam, and the ceaseless bobbing of his boat. With snow in the background, the "dark-blue and angry waves" contrasted nicely "with the white winter landscape."[23]

ONE "CANNOT SAIL OR ROW OVER THIS WATERY WILDERNESS without sharing the excitement of this element. Our sail draws so strongly that we cut through the great waves without feeling them. And all around, half a mile or a mile distant, looking over this blue foreground, I see the bare and peculiarly neat, clean-washed, and bright russet hills reflecting the bright light . . . from an infinite number of dry blades of withered grass."[24]

Spring water was usually muddy water, which "tossed" Thoreau and Channing "upon a sea of which one half is liquid clay, the other liquid indigo. . . . Such are the blessed and fairy isles we sail to!" Waves were polychromatic: "The water a dull slate-color and waves running high,—a dirty yellow where they break,—and long streaks of white foam, six or eight feet apart, stretching north and south between Concord and Bedford,—without end." When the mud settled out, they found an "undulating blue plain" of lively "tossing blue waves for a mile or more eastward and northward"; as Henry wrote, "A new season has come." In any month, he thought, "the tumult is exciting."[25]

In March "it blows so hard that you walk aslant against the wind." On the water "you scud before the wind in your tight bark and listen to the surge of the great waves sporting around you, while you hold the steering-oar and your mast bends to the gale and you stow all your ballast to windward. The crisped sound of surging waves that rock you, that ceaseless roll and gambol, and ever and anon break into your boat."[26]

During one "violent northeast storm" in April, "the meadows are higher, more wild and angry, and the waves run higher with more white to their caps than before this year. This wind, too, keeps the water in the river" by blowing upstream from the north. "It is worth the while to hear the surging of the waves and their gurgling under the stern, and to feel the great billows toss us, with their foaming yellowish crests. The world is not aware what an extensive navigation is now possible on our overflowed fresh meadows. It is more interesting and fuller of life than the sea bays and permanent ponds."[27]

In May the power remained strong: "I found myself in quite a sea with a smacking wind directly aft. I felt no little exhilaration, mingled with a slight awe, as I drove before this strong wind over the black-backed waves I judged to be at least twenty inches or two feet high, cutting through them, and heard their surging and felt them toss me. I was even obliged to head across them and not get into their troughs, for then I could hardly keep my legs. They were crested with a dirty white foam and were ten or twelve feet from crest to crest. They were so black,—as no sea I have seen,—large and powerful, and made such a roaring around me, that I could not but regard them as gambolling monsters of the deep."[28]

During July storms, "all farmers are anxious to get their meadow-hay as soon as possible for fear the river will rise." Indeed, a flooded river brought economic devastation. Hence a successful hay crop was "celebrated not with fireworks but a long sigh, that the hay was safely in." Thoreau witnessed incomplete harvests: "I see much hay floating, and two or three cocks, quite black, carried round and round in a great eddy by the side of the stream, which will ere long be released and continue their voyage down-stream."[29]

A surprise October storm brought the following waterfront scene: "A half-dozen boats at the landing were full, and the waves beating over them. It was hard work getting at and hauling up and emptying mine. It was a rod and a half from the water's edge," submerged by the rapid rise. This was the time to "look out for your rails and other fencing-stuff and loose lumber, lest it be floated off." Such storms were a repeat of spring floods, when: "All sorts of lumber is afloat—rails—planks & timber &c which the unthrifty neglected to secure—now change hands."[30]

October and November floods could be as strong as those of spring. "The reign of water now begins, and how it gambols and revels! . . . How they [the waves] run and leap in great droves, deriving new excitement from each other! Schools of porpoises and blackfish are only more animated waves and have acquired the gait and game of the sea itself. The high wind and the dashing waves are very inspiriting." In November "we rowed against a very powerful wind, sometimes scarcely making any headway. It was with difficulty often that we moved our paddles through the air for a new stroke. . . . We had to turn our oars edgewise to it. . . . There was quite a sea running on the lee shore,—broad black waves with white crests, which made our boat toss very pleasantly," with a wave height between trough and crest of "fifteen inches." During night sailing they could move "rapidly but mysteriously over the black waves, black as ink and dotted with round foam-spots with a long moonlight sheen on one side."[31]

Early winter added the element of ice. "Dark waves are chasing each other across the river from northwest to southeast and breaking the edge of the snow ice," breaking it up into "what arctic voyagers call 'brash,'" fragments of ice that "carry forward the undulation." The sound of brash ice is like that of static.[32]

IN THE STORMY MOOD OF THE FLOODED MEADOWS, Henry's attention was drawn to the water itself and to the excitement of running waves. But when the wind died down, the frothy bubbles burst, the sediment settled, and the water clarified, "a new feature" was "added to the landscape, and that is expanses and reaches of blue water." Musketaquid became "a succession of bays . . . a chain of lakes, an endlessly scalloped shore, rounding wood and field." Henry found it "far handsomer and more abounding in soft and beautiful contrasts than a merely broad river would be." He wrote, "The Great Meadows are . . . expanded to a large lake . . . a scene worth many such voyages to see."[33]

In this tranquil mood, his attention shifted outward to shoreline reflections. Five miles of open water could be seen in a single reflection of the Carlisle reach when the viewer was located at the right angle. Such water was beautifully "smooth and full of reflections," and colored "with a far deeper and more exciting blue than the heavens." Fair Haven Pond became "Fair Haven Lake, undistinguishable from fallen sky," he observed. "The great phenomenon these days is the sparkling blue water,—a richer blue than the sky ever is." When the breeze picked up, he noted, "the flooded meadows are ripple lakes on a large scale. . . . [A] copious living and sparkling blue water of various shades. It is more dashing, rippling, sparkling, living, this windy but clear day; never smooth, but ever varying in its degree of motion and depth of blue as the wind is more or less strong, rising and falling." Looking at the blue lakes through the edge of the forest created "a thousand little vistas," an "intimate mingling of wood and water." Using "our imagination," we "may navigate" those parts of the waters that were "concealed."[34]

His attention also shifted downward through transparent water for a view like that through a glass-bottomed boat. In another example of wildness reasserting itself, Channing and Thoreau saw "fields of potatoes and rye beneath still waters." They could "paddle right over . . . cow-slips in full bloom; their lustre dimmed, they look up with tearful faces." A clutch of submerged daffodils might hide minnows or their predator, the pickerel. Low roads that had been dry for the teamster the previous week became even smoother highways for a boat being rowed across them.[35]

"This dark-blue water is the more interesting because it is not a permanent feature in the landscape," he observed. Its spontaneity was its wildness. "There is the magic of lakes that come and go. The lake or bay is not an institution, but a phenomenon. You plainly see that it is so much water poured into the hollows of the earth." Under normal circumstances, "we are stiff and set in our geography—because the level of water is comparatively, or within short periods, unchangeable—We look only in the sea for islands & continents & their varieties." But here in Concord he was able to look forward to "more subtle & invisible & fluctuating floods—which island this or that part of the earth—whose geography has never been mapped."[36]

"My eyes are attracted to the level line where the water meets the hills now, in time of flood, converting that place into a virgin, or temporary shore," he wrote. "Ball's Hill and the rest are deep sunk in the flood. The level water-line appears to best advantage when it appears thus to cut the trees and hills. . . . No permanent shore gives you this pleasure." "There is no strand,— nothing worn; . . . It does not beat, but simply laves the hills." In places the strand was "so reddened with cranberries that I perceive them fifteen rods off, tingeing it." Like a meniscus in a test tube, it "kisses" the shore tremulously, creating an "inexpressibly soft curving line." This is "a shore where there are no shore marks," one that is "abrupt & surprising." Tributary valleys entering the lakes became "deep and narrow 'fjords.' "[37]

The line of flotsam from the spring freshet became a marker in time, "an endless meandering light-brown line, further from or nearer to the river." He loved "to see it even in midsummer," because it reminded him "of the floods and the windy days of the fall and spring, of ducks and geese and gulls, of the raw and gusty days which I have spent on the then wilderness of water, of the origin of things, as it were, when water was a prevailing element." In contrast to that tumult, there was now this tranquility.[38]

Reflections changed with the seasons. In April or November, the meadow lakes become "reservoirs of dark indigo amid the general russet and reddish-brown and gray." Between May and August the sea could be of many colors, with islands of "green hills rising out of it." "Dark-blue, clay, slate, and light-blue, as you stand with regard to the sun. With the sun high on one side it is a dirty or clayey slate; directly in front, covered with silvery sparkles far to the right or north, dark-blue; farther to the southwest, light-blue." Fall brought autumnal tints. Winter brought white reflections. "Methinks this rise of the waters must affect every thought and deed in the town. . . . I trust there will appear in this Journal some flow, some gradual filling of the springs and raising of the streams, that the accumulating grists may be ground. A story is a new, and in some respects more active, life in Nature."[39]

Eventually the transient lakes of every freshet drained away. Just before this happened, the meadows would be covered with a film of water too shallow to boat in. At such times he enjoyed the illusion of walking on water. "Went barefoot some two miles through the cut-grass," he wrote in 1856.[40]

THOREAU'S INLAND SEAS SHARED much with their oceanic counterparts. "I observe the phenomena of the seashore by our riverside, now that there is quite a sea on it and the meadow." The surface of the sea "has that same streaked look that our meadows have in a gale," and the waves break "with violence on this shore, as on a sea-beach." There was a rhythmic beat to the waves as well. "As on the sea beach, the waves are not equally high and do not break with an equally loud roar on the shore; there is an interval of four or five or half a dozen waves between the larger ones." Like the sea, the meadows were so broad that he could trace the shadows of the clouds over them. And as with the sea, the sound of water washing over the prows was the same.[41]

Land ho! "This first sight of the bare tawny and russet earth . . . over the meadow flood in the spring, affects me as the first glimpse of land" seen by a seaman "who has not seen it a long time." "After my winter voyage I begin to smell the land." In some Concord strands he found "the beach plum . . . the sight of it suggests that we are near the seacoast, that even our sands are in some sense littoral,—or beaches." And "when our meadows are flooded in the spring and our river is chained to a sea, then the gulls, the sea birds, come up here to complete the scene." Finally, the falls of the Concord River below the Fordway had the surf-clinging "rock weed of our river . . . It seemed as if our river had there for a moment anticipated the sea—suffered a sea-change— mimicked the great ocean streams."[42]

The meadows resembled the sea not only when they were under water but also when submerged by grass. "As I wade through the middle of the meadows in sedge up to my middle . . . I feel a little as if caught by a rising tide on flats far from the shore, . . . cast away in the midst of the sea. It is a level sea of waving & rustling sedge about me. The grassy sea—. You feel somewhat as you would if you were standing in water at an equal distance from the shore." The hay cutters, when seen from afar, were "up to their shoulders in the grassy sea, almost lost in it." And when the wind blew, it caused the meadow grass to undulate "like waves of light and shade over the whole breadth of [the] land, like a low steam curling over it . . . waves of light and shade pursuing each other over the whole breadth of the landscape like waves hastening to break on a shore."[43]

Finally, Thoreau's inland sea was rich in littoral fauna. Freshwater mussels gave the taste of true marine clams: "For refreshment on these voyages we were compelled to drink the warm & muddy . . . river water out of a clam shell—so that it reminds you of a clam soup—taking many a sup—or else leaning over the side of the boat—while the other leans the other way to keep your balance, & often plunging your whole face in at that when the boat dips or the waves run."[44]

After the excitement he felt at being the steersman of a schooner in Cape Cod Bay in the summer of 1851, actual sea voyages seemed to impact him less. In contrast to his inland sailing adventures out of Port Concord, his recollections as a saltwater passenger are descriptive and matter-of-fact. On December 29, 1854, he visited Nantucket. Returning to Cape Cod, the vessel he was traveling on was dangerously enveloped in fog. On a day trip on January 27, 1856, he traveled with Rickertson "and his boys in the Steamer Eagle's Wing, with a crowd and band of music, to the northeast end of Naushon, 'Woods Hole,' some fifteen miles from New Bedford; about two hours going. . . . Returning, I caught sight of Gay Head and its lighthouse with my glass." On yet another trip, this one in January 1857, he "saw Boston Harbor frozen over (for some time). Reminded me of, I think, Parry's Winter Harbor, with vessels frozen in. Saw thousands on the ice, a stream of men reaching down to Fort Independence, where they were cutting a channel toward the city. Ice said to reach fourteen miles. Snow untracked on many decks."[45]

7 | River Sojourns

THOREAU'S GENIUS FOR EXPLORING HIS WATERWAYS is most compelling for the stable decade beginning with his shift toward science in late 1850 and ending with the onset of his fatal illness in late 1860. The broad crest of this wondrous decade was the three-year interval between his late summer (August) 1856 epiphany about wildness and the late winter (February) 1859 death of his father, after which Henry became the official head of his family's household and CEO of the family business.

Thoreau's youthful excursions, explorations, tragedies, and unrequited loves were behind him. The protracted legal fight over Billerica dam and the divisive political angst of John Brown's raid on Harper's Ferry, Virginia lay ahead. His vocation as a land surveyor was providing a reliable income. His profession was going well: "to be always on the alert to find God in nature, to know his lurking-places, to attend all the oratorios, the operas, in nature." His journal was very systematic and rich with sojourning observations. *Walden* was slowly gaining attention. And for the first time in more than a decade, he had no debt to pay, no major project diverting his attention, no great excursions planned, and no great personal stressors.[1]

The zenith of that three-year crest arrived during the summer of 1858. That's when Thoreau retrofitted his boat to make it faster and stronger. Its original Spanish brown color was gone, replaced by a "pale green & whitish" color from bow to stern. This helped camouflage it during the walking parts of his fluvial and riparian excursions. He added power by splicing a new sail onto his old one, making a composite squarish sail, some five feet by six. "It was a "simple homely square sail, all for use not show, low and broad!" To raise it, he added rigging: "Before, my sail was so small that I was wont to raise the mast with the sail on it ready set." Now he was thrilled with the ad hoc conversion. "I like it much. It pulls like an ox. . . . The yard goes about with a pleasant force, almost enough, I would fain imagine, to knock me overboard."

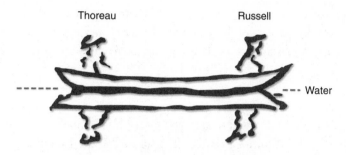

Thoreau   Russell

Water

FIGURE 18.   Double shadow. Redrawing and labeling of
sketch of double shadow of Thoreau's boat (on the bank
and the bed) from his journal of August 16, 1854. Boating
companions John Russell and Henry David Thoreau are
both wearing hats.

His "voyages" became faster and longer. He began to worship this new power
in a "heathen fashion," as if it were a "Grecian god." His boat became the plow,
drawn by a "winged bull," that cultivated his life on the river. These were years
of wonder and great joy.[2]

This wondrous decade was the most stable and productive of Thoreau's
life. He had resolved to remain a bachelor, to restrict his focus to Concord and
vicinity, and to live with his family in the Yellow House. His younger sister,
Sophia, also unmarried, became his companion. His third-floor sanctum gave
him a large and exceptionally private place to work. His boat place lay just
across the street, about one minute from his door, giving him almost instant
access to the water. The railroad station was around the corner, less than
three minutes away, giving him convenient daily access to the intellectual resources of Harvard College in Cambridge and the Natural History Society
in Boston, both of which could be reached within an hour. There he had un-
restricted access to excellent libraries, comparative collections, and knowl-
edgeable colleagues.

Once in this groove, Thoreau religiously followed a long-term, regular
pattern of reading in the evenings, writing up his thoughts in the mornings, and
sojourning in the afternoons, often on two or more of his three rivers. This
pattern was syncopated with the microclimate of his attic garret, which became too hot to work in during summer afternoons, and which was nicely

warmed during winter mornings by the chimneys. This work was coordinated with the seasons of the year, which offered very different possibilities for boating, depending on the daily conditions of wind and rain.

THOREAU MADE A CONSCIOUS DECISION TO GO DEEP into nature instead of broad. His goal was to acquire an exquisitely detailed knowledge of his particular valley, rather than a general knowledge about a wider geography. Though he traveled widely within the region, his focus was consistently local. By the spring of 1857 he had developed "a strong and imperishable attachment to a particular scenery, or to the whole of nature." In context, I translate "attachment" to mean a powerful sense of place, and "the whole of nature" to mean everything within his watershed. "All nature is my bride," he wrote. By staying local, he could see all "those familiar features, that large type, with which all my life is associated, unchanged." To go deeply was to "take the shortest way round and stay at home. A man dwells in his native valley like a corolla in its calyx, like an acorn in its cup. Here, of course, is all that you love, all that you expect, all that you are. Here is your bride elect, as close to you as she can be got. Here is all the best and all the worst you can imagine. What more do you want?"[3]

This insight—this declaration of love through understanding for his local watershed, and his commitment to know it scientifically—coincides with the crest of his river sojourning years, just a few months before upgrading his boat. By this stage in his life, Thoreau had become the valley's most highly respected local scientific authority, consulted by anyone with a question. He had learned to teach himself and then share what he had learned: "My knowledge now becomes communicable and grows by communication." Emerson joked in his journal that Concord should officially appoint Henry to that role and pay him for his services. Alternatively, he suggested that Henry open a consulting business and charge for his services. Eventually, the townsmen did just that: they hired him as a consulting river scientist.[4]

AFTER NINE WEEKS OF INTENSIVE RIVER INVESTIGATIONS during the 1859 portion of his river project, Thoreau's commitment to the watershed remained unchanged: "Why should we not stay at home? This is the land & we are the inhabitants so many travellers come to see. Why should we suffer ourselves to drift outside & lose all our advantages." "It is not the book of him who has travelled the farthest over the surface of the globe, but of him who has lived the deepest and been the most at home. . . . The poet has made the best roots in his native soil of any man, and is the hardest to transplant." To leave and study elsewhere would be to "excommunicate" oneself. "If a man is rich and strong anywhere, it must be on his native soil." "With the utmost industry, we cannot expect to know well an area more than six miles square, and yet we pretend to be travellers, to be acquainted with Siberia and Africa"—or even western New England.[5]

Two years earlier, when traveling through the Connecticut River Valley between Vermont and New Hampshire, Thoreau was "struck by the greater luxuriance of the same species of plants" in Concord relative to "up-country, though our soil is considered leaner. . . . Our river is much the most fertile in every sense. Up there is nothing but river-valley and hills. Here" in Concord "there is so much more that we have forgotten that we live in a valley." The axis of that valley was Musketaquid. And the place where Musketaquid met the Assabet at Egg Rock was the epicenter of his studies. This was as true for Thoreau as it had been for the Native Americans who preceded his culture.[6]

By the Numbers

It's widely appreciated that Thoreau kept a journal throughout his entire adult life. There were twenty-four years between the first entry, on October 22, 1837, and the last, on November 3, 1861. Less well appreciated is that the vast majority of its words—probably more than 90 percent—were written during his wondrous decade of stability. Even less well appreciated is that about half of the total word count comes from the five-year period when Thoreau's chief work was being a scientific naturalist. By then, his journal had mainly become a repository for what he called "Field Notes." On March 21, 1853, he contemplated using this as a title for his journal, which recorded his professional, though unpaid, work.

Where did Henry go on his sojourns? With whom? What did he do? These sound like simple questions. And they are, if we're talking about a day, week, month, or a season. But what interested me for this book project was the whole slice of his life as a regularly sojourning scientist. I knew that the only good answers would come from making tallies of what, where, when, and with whom, followed by a search for broad patterns using statistical methods. Very quickly I learned two things. The first was that this decade was indeed a remarkably stable period, one regular enough to yield meaningful generalizations based on journal entry frequencies. The exact duration I reconciled ranged from May 12, 1850, through June 7, 1861. The total was 3,653 days spread over ten calendar years, three of which were leap years. Though the journal shows great changes in content and mood during this decade of consistent observations, the overall pattern for entries remained much the same, making proportional comparisons possible.

The second thing I quickly learned was that his descriptions of where he went specifically, with whom, and what he did were too inconsistent to meet any standard of statistical rigor. The journal was private, so there was no need to tell himself things he already knew. For example, the pronoun "we" doesn't identify whom he was with. Though I gathered data for sixty-two categories of information from 6,958 journal entries, my research method quickly narrowed to four nested geographic questions: Was Thoreau in Concord that day? If so, did he sojourn beyond the town streets? If so, did he go to the river, or to some interior landscape? And if the latter, did he go to Walden Pond?

Specifically, I wanted to test my hypothesis that Thoreau was more a man of the river than of the interior woods, meaning more of a boatman than a woodsman. A corollary would be that the link between Thoreau and Walden Pond is stronger in our minds than it is in his journal.[7]

By my count, of these 3,653 days, Henry made journal entries for 2,994 of them, or 82 percent of the time. Of those entries, 2,735, or 92 percent, were written when he was in Concord. Of those Concord entries, 2,121, or 71 percent, involved a local sojourning experience beyond the town center. Of those local sojourning experiences, 1,466, or 69 percent, were to one of his three rivers. This means that, based on the daily frequency of his visits, river sojourns were more than twice as common as inland sojourns. Unequivocally, Thoreau's default destination was his system of three-rivers, his go-to place for finding God in nature, most often by boat. In contrast, only 281 of the 1,466 local sojourns were to Walden Pond and its nearby woods, making visits there one-sixth as frequent as visits to the rivers. Walden visits peaked in 1852, when he was researching this locale for new material to add to his old *Walden* manuscript, and in 1856, when he began visiting this floristically unusual place for his new program of systematic phenology.

For the full ten consecutive sojourning years between 1851 and 1861, he wrote an average of 300 dated journal entries per year. Of these, an average of 274 were penned in Concord, of which 212 involved local sojourns. And of those local sojourns, on average 147 involved the river, with only 28 involving Walden Pond. After his transitional year of 1851, he became very consistent with his sojourning and journaling activities. For the succeeding nine years through 1860, he made 311, 313, 304, 300, 316, 315, 323, 302, and 313 entries, of which 233, 203, 243, 212, 226, 200, 236, 225, and 224 were local sojourns. The year-to-year stability of these numbers is extraordinary, with significant year-to-year variation showing up only for specific activities. Here is someone who, after fourteen years of post-college trial and error (1837–1851), had finally found his groove and stayed in it until pushed out by fatal illness.

One of the most interesting observations from this data set involves 1854, the year of *Walden*'s publication (the manuscript was submitted on March 13, during the spring freshet). Henry made more sojourns to the river in this year than in any other: 204 in total, compared with the ten-year average of 147. River visits for the next two years were above the average, with 173 and 151. The first full year after Walden's publication, 1855, was the most lopsided year for Henry's attention to the river, relative to Walden Pond. That year, he journaled about the river nearly twelve times more often than about the pond, a ratio twice the long-term average. Based on these three ratios, I interpret that, with the lingering demands of the *Walden* manuscript and the anxiety of its launch finally behind him, Henry's attention was drawn back to his first love, the river, even more exclusively.

I also kept track of the number of days Thoreau went boating on the river, rather than simply visiting it. Keep in mind that boating is possible only about nine months out of the year, and that the tally I present is a minimum value; my conservative criterion for assigning a boating day was an explicit mention of the word "boat" or text indicating that a boat's presence was required. Nevertheless, the minimum frequency of boating during this ten-year stretch is forty-four days per year. Though this may not seem like much, it's four times more frequent than the daily boat use for registered owners of small watercraft in the United States today. Explicit boating sojourns constitute approximately one-third of the total number of Thoreau's river sojourns. Notably, the minimum frequency for boating is nearly double that of visits to Walden Pond.[8]

THE MAIN OUTCOME OF THIS ANALYSIS was a spreadsheet matrix that allowed me to sort his river-related "field notes" by season, place, and topic. To share this with you, I adopted Thoreau's model of the Kalendar, which converts a decade's worth of observations into a single representative year. Instead of four seasons or two solstices, I found ten discrete river moods that Thoreau identified using discrete (though often overlapping) criteria. Two of these have already been described: the stormy gray and tranquil blue versions of his transient flood lakes. My representative year is dominated by entries from 1856–1858, which puts it right in line with the narrative sequence of this book.

My purpose is to put you, the reader, into the boat with Thoreau under the full range of conditions that he saw, heard, and felt during his river sojourning years. My über-goal is to convey his sense of place, which rays out from the triple point T-junction at Egg Rock. My initial draft of this section was a thirty-five-page manuscript dominated by quoted descriptions. As this book took shape, however, I decided to boil this section down to seven manuscript pages, which required dense paraphrasing. As you read, keep in mind that the vast majority of his observations were of a watershed he knew was being pervasively changed and locally overwhelmed by the human agency.

On the Water

The end of every Musketaquid winter was a festival of ice, usually seen from the nearest bridge because boating was too dangerous. Jagged floes two feet thick and tens of feet across circled in great eddies before being broken up and thrust edgewise between the piers. Audible crunches and cracks rose to a roar. Striking bridge abutments or each other, floes would shatter into fragments, or thud with seismic impact. Those that made it through the piers floated high on the water, colliding with whatever lay in their path. Bankside trees had their bark rubbed off. Some were uprooted like matchsticks, their branches broken away and their trunks used as battering rams to break up the next bridge downstream. Tufts of meadow sod, torn up by the ice, gathered together to create earthen dams, or were dispersed to litter downstream meadows.

On ponds such as Walden, the transition from frozen to unfrozen conditions was largely a silent process of melting in place. There Thoreau went to bed with "cold gray ice" on the surface and awoke to a "transparent pond." In comparison, Thoreau's three rivers behaved like young stallions bursting their "icy fetters." The bucking up of the ice was driven by pressure from below. That same warm sun that melted slab ice on Walden also melted the snowpack throughout the watershed. Trickling downward through the soil, liquid water was shunted sideways toward the river in aquifers, where it seeped into the channel at depth and physically lifted the ribbon of river ice upward. The result was a pair of rift zones between land-fast ice and uplifted ice, from which leaked and gushed pressurized water stained yellow by the tea of dissolved organics. At some threshold height, the ribbon of ice would disintegrate into angular floes up to a hundred feet wide that traveled together like cattle in a stampede. Down they went, tipping on edge, swirling in whirlpools, thrusting one above the other, and battering whatever was in their way, especially the wooden built environment of piers, docks and boathouses. When the stage was high enough, the stampede overtopped channel levees and diffused over the meadows, where they might be reconcentrated by the wind. The result was often a "very wild and arctic scene."[9]

River breakup in Thoreau's day was rendered more violent by the human makeover. Forest clearing increased the strength of cold winter winds, which swept the river clear of snow, allowing more heat to be extracted from the water. The result was thicker ice. To this effect was deeper freezing caused by more sluggish flows associated with the Billerica dam and the upstream bars it raised. Forest clearing also accelerated the snowmelt, lifting this thicker ice more quickly. The construction of bridges and causeways concentrated and amplified this force.

Thoreau was sometimes so impatient to begin his boating season that he poled his way through the ice of Fairhaven Pond to get upriver. On one occasion Channing and Thoreau physically broke up a large floe so they could squeeze beneath the stone arches of the railroad bridge. They did so by "lying flat and pressing our boat down" into the icy water. Once on the Assabet, Thoreau and Hawthorne rode a floe downstream while towing their boat behind for safety's sake. After the channel was cleared of ice, the high water and high winds of breakup combined to create the most invigorating sailing experiences of the year, especially on the Great Meadow. Some April gales were so strong that Thoreau didn't have the strength to raise his rigging, and the waves were so high they swamped his boat with ice water.[10]

EACH SPRING FRESHET LEFT A BIRTHMARK for the boating year to come, a line of wave-dashed flotsam and dried foam at the highest stage. Each was a miniature beach composed of broken twigs and branches snapped by breakaway ice and pulverized by grinding floes. Anything that floated gathered to create flotsam shorelines durable enough to last all year. They reminded the hay-makers that they worked on the bottom of a drained lake where the water had been head-high just a few months earlier. Nine feet was the vertical range in stage for this river.

Eventually the spring freshet drained back into the main channel beneath banks not yet leafed out. The default color of the banks was brown, though gray and russet were common. The "clear, placid, silvery water" was broader than during summer, because the edges weren't yet topped by lily pads, crowded by leafy overhangs, and shallowed by aquatic weeds.

With every inch of drop in the water level, the banks got reciprocally higher, increasing Thoreau's privacy. Fresh mud along the shore displayed the tracks of aquatic mammals such as minks and birds such as herons. This ri-parian spring was a time of great contrasts. Cold winds blowing over snow country to the north chilled him to the bone, while south-facing coves were warmed by the strengthening sun. Gusts blowing over the gallery forests flat-tened regular ripples with dark "tumultuous" blotches. Others whirled dried leaves from the previous autumn to great heights. His boat was dusted with grit blown from windswept fields.[11]

Gradually the physicality of late winter gave way to the return of life. In response, the philosophical tone of Thoreau's journal was overshadowed by a month-long rush of biological "field notes." His common practice was to slowly drift through riparian habitats being refilled with species. The shrill chorus of the thousands of spring peepers gathered for their mating orgy was an early sign. Larger wood frogs rang out later, and even bigger bullfrogs croaked later than that. Yellow salamanders, which he called lizards, wriggled about in search of each other. Sunfish, pickerel, and perch moved up from deep water to sandy shallows so they might absorb the warmth of the sun. When sailing strongly and steadily, Thoreau watched bright reflections and heard loud splashings as "whole schools of fishes" leapt out of the water ahead of his boat. Muskrats dove for clams with renewed vigor. Hundreds of turtles tumbled from rustling banks into the cold water as his boat approached. Pussy willows and alder catkins put on a great show. Stinking skunk cabbages were visited by bees long before the bass trees flowered.[12]

Twin migrations crisscrossed. From the south came the birds. High above were those in transit to northern climes, particularly the geese and ducks, which merely rested on Musketaquid waters. Smaller neotropical birds came and stayed, filling the riverbanks with birdsong long before the interior woods. From a million dormant plant stems came the general "quire" of spring. Robins cheeped in otherwise cheerless weather. From the north came the eels and suckers, migrating upriver from the sluggish lower reaches to the higher brooks to spawn.[13]

River vegetation leafed out like a green embrace from above, below, and the sides. Slowly the mottled shade of overhanging branches shadowed the river edges. The dust of pollen, the confetti of burst buds, and the downy fluff of the black willow coated the stream so thickly that a wake of clear water trailed his passing boat. Rising from below were the aquatic plants creating the local version of a kelp forest: polygonums, potamogetons, and pontederias. Crowding the sides were horsetails, flags, and, most important, the lilies, yellow first, followed by white.

AS THE RIPPLING RIVER SURFACE CONTINUED to drop in stage, the spring birdsong gradually faded. The noisy urgency of mating and territorial defense gave way to the quieter brooding of eggs. The crescendo of insect noise rose to its peak: crickets, bees, and mosquitoes, chirping, buzzing, and humming, respectively. Dragonflies, butterflies, and a myriad of others were noiseless. Thoreau recalled one June sunset with special delight. It was shortly after dusk. Thousands of frogs were croaking a deafening roar. A low mist hung above the water. In that direction, the sparkle of the planet Venus was multiplied by countless ripples, the light from each reflection haloed by the mist. Meanwhile, above the adjacent meadows the "coppery light of fireflies" was multiplying as dusk gave way to dark. This was a river summer scene that no forest or field could match.[14]

In June the "shad-flies" or "ephemerae" came out in such numbers as to darken the summer skies. A snowlike flurry of insects coated freshly painted houses in fluff. On the water, a chaos of fish leapt at them from below as swallows dove down at them from above. On such days, Thoreau would paddle to Clamshell Bank to watch these aerial dogfights.[15]

Summer announced itself with rumbles of thunder and the beginning of strong convective storms. At any moment boating could be interrupted by a squall of thunder, lightning, hail, and "coarse hard rain." Thoreau and

Channing would land, invert their boat, and take refuge on the "cat briars." When traveling alone on the Sudbury, Thoreau was sometimes forced to find shelter inside a cave near Lee's Cliff. From it he saw the "forked flashes" and heard the "hollow roars" of the storm.[16]

Summer announced itself with a switch from sailing to rowing, at least as a default mode. The regional stillness and heat of summer, the drop in stage, the growth of edge plants, and the loss of thermal contrasts between land and water confined him to a narrower, more windless channel. During one sluggish sail he counted ten dragonflies resting on his canvas. Boat seats became "intolerably hot."[17]

With summer came the onset of swimming, which Thoreau called "bathing." Each year he renaturalized as an amphibious creature to inhabit the denser habitat of the river otter. He took to the water to cool down, and then to sunny sandbars to bask. Sultry, furnace-like, and gloriously hot weather invited him into an enveloping embrace he never felt otherwise. All of his favorite swimming holes and basking bars had been enhanced by the higher water and stronger flows of his epoch, relative to those of the Holocene.[18]

Torrid summer heat offered Thoreau the experience of "water walks." When it was too hot to sojourn above the water, he did so within the water, walking the dustless road of the Assabet with more luxury than a "Roman Emperor." He could walk down the middle of the river on the same path he would skate over in winter. Naked except for his hat, the minnows and sunfish following him like a Pied Piper, he walked over a bottom that varied from ankle-deep sandy riffles to neck-deep squishy pools. Rust-stained ledges of bog iron in the bed signaled powerful erosion during his era. Elsewhere the bed was felted with sawdust, signaling scour holes being filled by sediment pollution from upstream factories. The process was a bit like boating without the boat.[19]

In early summer the tortoises locked themselves together in coital embraces that lasted for hours. By midsummer they had migrated to exposed sandbanks to lay their eggs. The abundance of sand hills associated with land clearing likely improved their habitat. In the river Thoreau would sometimes pass large snapping turtles, drifting downstream with their heads sticking up like periscopes. When transient high water was subsiding on the meadows, he found snappers crawling on the bottom by the dozen. On one occasion he stripped off his coat, rolled up his shirtsleeves, leaned over the edge, and plunged his arm down to grab a big one by the tail, hauling it up into the boat. For a while Thoreau and the turtle faced off. What impressed him most was the turtle's violent primitiveness. The namesake "snap" struck him as more auto-mechanical than volitional-biological.[20]

The fragrant white water lily was the crowning achievement of aquatic summer, equivalent to the rose of the terrestrial realm. The rose has its thorns, and the lily its fetid muck, from which great beauty emerged. Thoreau took one boating excursion before dawn just to hear the lilies pop open.[21]

On the meadows, the grass, sedges, pipes, and herbs grew like green flames. In midsummer the haymakers would harvest them for cattle fodder. During the "vintage" of haymaking, Thoreau became proud of his furtiveness: "Large as my sail is, it being low I can scud down for miles through the very meadows in which dozens of haymakers are at work, and they may not detect me."[22]

Herons, bitterns, cranes, and stake-drivers stalked the summer meadows, searching for hapless aquatic creatures, especially fish and amphibians. Sunfish guarded their underwater nests, bullheads swam with their broods, shiners gathered in schools, and pickerel lay in ambush for creatures that came too close. Hatched birds were everywhere, and their parents were scolding anyone and anything that drew near.

Late July and August brought dog-day weather, a blue haze, with thick and mildewy air, and an orange-red sun, greatly engorged as it rose and set over the water. There were scorching heat waves, with the temperature in the shade on Main Street nearing but never reaching triple digits. With no wind, sailing was out of the question. Rowing raised a salty sweat, and he required repeated dunkings to stay cool. Sometimes Channing and Henry leaned over the side simultaneously to counterweight each other, plunging their heads into the water. In shallow sluggish reaches, the water felt almost as warm as the air. There the water developed a bluish, nearly imperceptible, and sometimes slightly iridescent color with an almost oily surface; from a distance it resembled the color of slate. Copious springs draining into it from the sandy glacial aquifers on either side kept the main channel from going dry. During this season, Musketaquid became a chain of slackwater lagoons separated by congestions of weeds.

DROUGHT WAS BAD NEWS. Strong evapotranspiration over the watershed's hundreds of square miles sucked moisture from the ground. Brooks that gurgled at night when the trees were asleep dried up each afternoon when they were drinking deeply. Eventually many brooks went quiet. Millers shut down their mills. Lily pads sagged down in twisted, filamentous heaps to rot in the baking sun. The river took on the sulfurous smell of low tide.[23]

Desiccated meadows brought the smell of carrion to Concord. On the surface were millions of small pools, each where tufts had been lifted up by the ice, leaving a divot. In them, Henry found thousands of young fish—breams, pouts, pickerel, minnows—gathered together to die in dense concentrations. The result was rotten fish, mixed with the rotten-egg odor of hydrogen sulfide being released from black muck. Dust from dry meadows combined with dust from farm fields and dust from roads to whiten the lily pads for more than a mile on either side of Concord Center. Smoke from deliberately set fires added to the haze of hot summer. Some farmers took advantage of low water to burn the meadows free of incipient shrubs, believing it would improve them. Instead, the meadows often burned like underground coal mines, smoldering below the surface, lowering the land, and increasing the wetness when normal conditions returned.

Dry meadows offered a bonanza to the stalking birds. Eating from the shrinking pools was like forking fish from a pie plate. On one August day Henry boated by thirteen great blue herons spaced out along the river, each feasting on its own patch of concentrated meadow life. In one case he saw through his telescope glass what he thought was a camouflaged fisherman. Drawing nearer, he decided at first that it was his friend Channing, toweling off after a swim; then he perceived it as a "maiden in a bathing dress"; finally he saw it to be a blue heron pluming itself.[24]

THE MOST DELIGHTFUL SAILING CAME IN EARLY AUTUMN. The stifling heat of summer stillness was blown away. Subtropical storms created transient blue-water lakes. "The winds of autumn begin to blow." Once again Thoreau watched his "mast bend in these safe waters." His boat cut through a cleaner liquid, no longer dusty and unctuous. No longer did his oars and his rudder leave bubbles that lasted an hour or more. Instead, they broke quickly and cleanly, the water being "less adhesive," or more elemental. The water also became more transparent beneath his boat, as if "clarified by the white of an egg or lime," making the river resemble an aquarium. This was partly an optical effect, due to the whitening of the sky with increased haze during the final days of late summer. Another cause of clearing was the reduction of dissolved nutrients, which limited algal growth. Another was the dieback of the lush growth of aquatic weeds. They became what Thoreau called "in-browned," rotting in place until they softened. Leaf by leaf, plant tissues detached and floated downstream.[25]

To these in-stream effects were added those of the watershed. Evapotranspiration was shutting down, and stream flow strengthened. The rains came back as a consequence of regional weather. The winds picked up to blow the in-browned weeds to the shore. The boatman's attention shifted downward into the living waters, where a whole new world opened up below. Paraphrasing *Walden:* Heaven is beneath our boats, as well as over our heads.[26]

THE RIVER'S FIRST AUTUMN was an aquatic phenomenon: a clarification, cooling, freshening, and widening of the river, accompanied by the dieback of submerged and shoreline plants. Its second autumn was riparian: a blaze of bank colors and a fluttering fall of leaves made twice as beautiful as on land, given the silvery blue reflections involved. During this bankside autumn Thoreau had to clear his boat of golden willow leaves—leaves destined not for soil humus but for a journey downstream toward the sea—before he could go out onto the water to enjoy some of the most glorious days of the year.

The red maple was the first to raise its "scarlet flag" on the water. The willows blazed yellow, combining with the maples to paint the bank the color of flame. The buttonbush went from green to dull brown to naked twigs. Cruising the river on successive days was like watching a slow-motion kaleidoscope of color. The larger, overhanging trees of the riparian woodland—white maples, walnuts, oaks, and other bankside species—dropped their waxy, curled leaves onto the water, where they became "rude boats" and "leafy skiffs" that floated and sailed about by the millions. In places they gathered in fleets so dense they covered the water completely. Gradually they softened and sank into the water, having been wetted above by rains and dew. Down they went to become trapped in pools or behind the Billerica dam.[27]

The colder, fresher waters of late fall gave the river a "singularly ethereal, celestial, or elysian look" when the light was right. The meadows above them lightened to an amber brown. Above the meadows were the darker browns of grain field stubble and dried pasture grass, and above them the hills took on a tawny brown color.[28]

Every animal put on its winter coat. Thoreau did so incrementally, adding a light coat, then gloves, then heavy pants, a heavy coat, and finally boots. By the end of August, sailing on windy days could be a chilling experience. Umbrellas became essential equipment, sometimes serving as an extra sail. The squirrels built their nests on the outer branches of tall trees. The muskrats constructed their conical lodges on the leafless shore of the river. The clams migrated downward to deeper water. The spiders released their gossamer threads to colonize downwind.[29]

This was driftwood season for Thoreau, who gathered great loads "of various kinds," weighting his boat so low in the water that it moved slowly and sluggishly, like the canal boats of yore. Once brought to harbor, the wood was off-loaded and carted across Main Street to his house, chopped in the yard, and hauled up to the garret to feed his iron stove.[30]

The rains become harder and colder. The river stage rose under gray skies and fell under clear skies as the weather patterns of later autumn did battle with each other. Life seemed to vacate the channel. Psychologically, this was a time of drawing inward. The shadbush gave him hope. While other plants were dressing down, it "puts forth fresh & tender leaves." Thoreau had "shad-bush thoughts," which were confidently full of faith and hope.[31]

BY MID-NOVEMBER THE PHYSICS OF COLD, clean beauty were ready to replace the luxuriousness of life during the humid, hazy, suffocating, panting heat of summer. This was a lull time in river activity. The leaves had flamed out; the hills turned russet, the meadows golden, and the river a Valhalla blue. The river was "waiting for ice." Thoreau now preferred rowing to sailing, if only to warm himself with the exercise. By late November, however, the water began to freeze on the paddles and numb his hands. By early December he had to break his boat out of the shore-fast ice to sojourn. Channing admired Thoreau's resistance to cold: he "voyaged about his river in December, the drops freezing on the oar, with a cheering song; pleased with the silvery chime of icicles against the stems of the button-bushes." One day, after a four-mile row downriver, they returned with a headwind so strong they arrived well after dark with their "feet and legs numb and cold with sitting and inactivity."[32]

In spite of cold temperatures, the aesthetics of freeze-up was a great incentive to go boating. Thoreau enjoyed rowing through paper-thin black ice so that he might hear the crack and tinkle of its fragments. He enjoyed rocking his boat and watching the undulations of the wave move beneath thin, flexible sheets as if they were rubber membranes. He particularly enjoyed the Assabet during these conditions, where the progression of freeze-up from the sides, combined with the storage and retention of water by millers, created a shelf of ice around every boulder, bank, and log, something like the bracket fungi of trees. Each thin shelf marked the stage of post-midnight cold, a marker for the water level nearly as sharp as sawdust on the bank.

But it was the glorious crystals of solid water that drew him out into the clear subarctic air. Thoreau found "shooting crystals," which were "single crystal spears up to six feet long, narrow and sharp." These he saw only under the right conditions of light, low in the sky. Each boat trip offered a view of uncollectable crystal shapes: rosettes, cobwebs, mosaics, gossamer threads, fern leaves, ostrich plumes, fir trees, fletched arrows, and graphic ice resembling ancient lettering. Where surface ice was broken up or rained upon and then refrozen, he found amorphous constellations ranging from milky slush to broken glass to a hobbled, dimpled surface. Some ice, he recognized, was slush that squirted out from fractures under artesian conditions to resemble the pillow lavas of undersea volcanoes, or bread risen out of the pan. When black ice thickened to three inches, it was solid enough to support his walking, even though it flexed up and down like a stiff trampoline. To see this threshold thickness required that the ice be cracked by milky-white fractures. When one reflects on the hundreds of descriptions of thin ice Thoreau gave us, it's almost a miracle that he survived to tell us what he saw.[33]

Freeze-up was a contest for the middle of the river, a tug-of-war between atmospheric cold attempting to close the channel and aqueous heat attempting to keep it liquid. This back-and-forth toggled between cold nights and warm days and, at a longer time scale, between weather patterns dominated by northerly and southerly winds. Thoreau enjoyed boating the river when it was a narrow canal between ice closing in from both sides. That's when the crystal frame of the ice became an intermediary between turbulent liquid and turbulent treetops. Eventually the contest for the middle was won by the cold, the boat was put in the cellar, and the boatman donned his skates.

Thoreau puzzled over the fact that there was only one type of liquid water but a myriad of frozen types. The weight of snow on river ice caused the ice to sink downward, saturating the snow. This froze into a homogeneous mass of gray-white translucent ice called snow-ice. The penetration of cold below that snow-ice added clear layers from the bottom, like the rings of a tree or the layers in a clamshell. River ice thus became a variegated sandwich of solid water that was infinitely interesting to Thoreau for its many forms and colors. In some places the river remained unfrozen even during the coldest weather, there being too much groundwater heat from springs or turbulent heat from flowing water, especially at bridge narrows. Once in place, ice immediately began to metamorphose in transparency, color, and hardness in response to heat and sunlight from above and flexure from the sides.

Freeze-up drove the majority of Concordians indoors. Furtive river animals took notice. The river otters that Thoreau tracked in the snow came out to enjoy the solitude. Like seals, they were exclusively aquatic, subsisting mainly on fish and being vulnerable away from the water. Like seals, they were highly social and enjoyed group play. Over the span of fifteen winters Thoreau had witnessed their spoor, tracks, and trails, and heard stories about their being trapped and hunted, but he never laid eyes on one. "A strange track in the snow" is a good summary of his many observations. In total, I count nine journal descriptions of otter behavior that Thoreau reconstructed based solely on the physical evidence left behind during this season.[34]

DEEP WINTER WAS THE ONLY RIVER mood without boating. Thoreau felt compensated by the improvement in pedestrian transportation. The bridges of the built environment became superfluous because ice bridges were everywhere. When the river was covered by new ice or when its rift zones were well developed, he could skate or walk up to sixty miles per day. When the landscape was covered by heavy snow, the river remained a good highway because the snow and ice merged into the snow-ice described above. Nor did the river channel commonly drift over with snow, either because there wasn't enough wind fetch for big drifts or because the snow was blown to one side or the other.

The situation in the adjacent meadows was different. There the wind fetch was up to two miles long, and fallen snow drifted into great waves that resembled those on the ocean in form but were stationary at the time scale of minutes. At the time scale of hours, however, they moved like dunes in the desert. Walking the meadows during a strong northerly wind was nearly impossible.[35]

One of the odd things about midwinter that Thoreau enjoyed was the symphony of sounds made by the ice: groaning, booming, farting, belching, cracking, crinkling, et cetera. Like the earth's crust, thick ice is a continuous rigid slab. Seemingly small cumulative stresses caused by changes in temperature, buoyancy, and atmospheric pressure created instantaneous strain releases, essentially brittle fractures—ice-quakes, if you like. The well-known line from *Walden* "Who would have suspected so large and cold and thick-skinned a thing to be so sensitive?" is actually about the river at Fairhaven Pond, a wide spot on the river.[36]

The deep freezes of winter's culmination sent most animals into a deep hibernation. The Great Meadow, Thoreau noted, formerly a "museum of animal and vegetable life," became a wind-drifted sea of sun-crusted snow with only a few plant stems rising up here and there. It was under these conditions that the wary foxes came out to forage, and Henry came out after them. Though the subaerial winter landscape was largely frozen into stillness, animal life remained active in the dark aquatic underworld beneath the ice. This was a tenuous life spent waiting for spring, when rushing waters and the growth of aquatic vegetation would once again infuse the water with the oxygen needed to sustain animal life.[37]

Above the ice, in dazzling sunlight, Henry would sometimes build a fire with "fat pine." There, in the middle of the channel, he watched the snow melting, the water sizzling, and the coals descending downward to darkness. Scarcely one month later he would be boating over that very same spot, and three months later swimming there with the snapping turtles and schools of silvery fish.

8 Consultant

THE THREE SUMMERS OF 1856, 1857, AND 1858 were a boatman's paradise. Deep summer floods each year created blue inland seas on which Thoreau sailed away. Unfortunately, they were a haymaker's nightmare, bringing Musketaquid farmers to an angry boil.

In August 1856 a subtropical soaker deeply flooded the meadows, ruining whatever hay was left standing. Based on rainfall data from nearby Waltham, almost fourteen inches of rain fell during that month. This was nearly twice as much as the next highest monthly total on record, and nearly three times the average. Thoreau commented in his journal about this heavy rainstorm, noting that the hay crop was a total loss: "I hear that a Captain Hurd, of Wayland or Sudbury, estimates the loss of river meadow-hay this season in those two towns on account of the freshet at twelve hundred tons." Within three years he would be working for that family from Wayland, and would learn how to spell their name: Heard.[1]

In May 1857 a late spring freshet dumped nearly five inches of rainfall on the watershed. Though too early to spoil the hay, it ruined all crops planted in low ground by drowning the seedlings. "Gardeners wish that their land had not been planted nor plowed," Thoreau observed on May 20. That year would see yet another lawsuit brought against the factory dam in Billerica by the meadowland proprietors. As usual, their case was dismissed for the century-old reason that the blame lay with the Commonwealth for passing such misguided legislation, rather than with the industrialists, who had broken no laws.

The final blow—literally—came in early July 1858. Intense blowing rains in late June dropped more than twice the month's average total precipitation on the valley. The timing was terrible, for the rain had come too early to ensure a good first crop and too late for there to be hope for a second crop. The plain facts, as recorded in the later Joint Committee Report, spoke for themselves: "The usual annual amount of hay secured from the aforesaid meadows, before they were flowed, was one thousand tons, while in the year 1858 the crop of hay secured from said meadows, did not amount to more than one hundred tons; and that the usual annual crop of cranberries, previous to the meadows being flowed, amounted to three thousand bushels, while the crop of cranberries in 1858 did not amount to ten bushels." These losses were 90 percent for hay and 99.6 percent for cranberries. Thoreau noted that under such circumstances, farmers with diversified land holdings were beginning to divest themselves of meadow agriculture: "to get rid of their river meadow now, since they can get more and very much better grass off their redeemed swamps, or meadows of their own making, near home." Those without access to swamp hay, however, were still coming "a good way for their meadow-grass, even from Lincoln."[2]

Thoreau poignantly captured the lingering sense of defeat these farmers were facing year after year: "I see an empty hay-team slowly crossing the river, in the shallowest place. . . . They have not got more than half the hay out of the meadows yet, and now they are so wet I see but one team here. Much grass will be lost." Ten days later he witnessed "men at work in the water a foot or more deep, saving the grass they had cut. . . . Many tons stand cocked up, blackened and lost, in the water, and probably they will not get the grass now standing."[3]

AFTER THREE INCREASINGLY BAD YEARS, the people of the meadowland rose up in protest. On the twenty-seventh of December, 1858, in the Town Hall of Concord, more than two hundred citizens gathered for a historic meeting. Was Thoreau among them? An account of the event noted, "The meeting was called to order at one o'clock by Simon Brown, Esq., of Concord, and, on motion of Samuel H. Rhoades, Esq., of Concord, a committee was appointed to report a list of officers. Simon Brown was chosen President, Col. David Heard, of Wayland, and seven others, Vice-Presidents, and Mr. R. F. Fuller, of Wayland, and Dr. Joseph Reynolds, of Concord, Secretaries." Brown urged them "to organize—to take a stand, and raise their colors and nail them to the mast. (Applause). From that moment onward, great agitation commenced."[4]

Later that winter, on February 10, 1859, the angry landowners of Sudbury voted unanimously that the "Selectmen of said Town be instructed to petition the legislature" for relief regarding their wet meadows. Specifically,

Henry Vose and 176 others wanted an act that would require removal of the compensating reservoirs above them, and the Billerica dam below. Two days later Concord joined them with a petition signed by Elijah Wood Jr., one of Thoreau's boating companions, and 178 others. Wayland followed with Richard Heard and 116 others, and then Bedford, with P. F. Chamberlain and 68 others. These first four proposals were referred by the House of Representatives to its Committee on Agriculture, with additional proposals following on February 26 and 28.[5]

The summary document for all towns was titled "Petition—of the Towns of Concord, Wayland, Sudbury, Bedford, and Carlisle, Praying for the Removal of Nuisances and Unauthorized Encroachments in Concord River—to the Honorable Senate and House of Representatives of the Commonwealth of Massachusetts." Clearly and eloquently it stated their plight:

> Your complainants and petitioners are chiefly agriculturists, and our most valuable lands, in respect of natural fertility and productiveness, were, from time immemorial, our meadow lands, situated on the banks of the Sudbury and Concord rivers, constituting an alluvial formation, which, without reckoning windings, reaches some twenty-five miles, through Wayland, Sudbury, Lincoln, Concord, Bedford, Carlisle, and Billerica, and comprises from eight to ten thousand acres. . . . [These lands] had always been subject to overflow in the Spring freshets, and in cause of deluging rains; occasionally, though rarely, at other times ; but these inundations, always leaving, like the Nile, a fertilizing deposit, and passing off without any important artificial obstruction, were the blessing of the husbandman.[6]

All together, 540 individual landowners joined in what was effectively a class-action lawsuit with statewide and national implications. Their principal legal advocate, the Hon. Judge Mellen, was explicit about the scope and importance: "The Petitioners represent the interests of thousands of meadow-owners in all parts of the Commonwealth." This was a tale of woe with two clear culprits. Most important was the state itself. Using capital letters for emphasis, Mellen argued: "THIS DEPRECIATION WAS WHOLLY CHARGEABLE TO THE MIDDLESEX CANAL CORPORATION," now defunct, which had built the Billerica dam at the outlet of the watershed and sold its water privileges to factory owners. The second culprit was the City of Boston, which in 1844 had

FIGURE 19. After ice breakup. Sudbury River looking south toward Fairhaven Hill. January 2012.

"purchased, and built, or repaired, two vast Reservoirs, distant from, and independent of," their water supply reservoir of Cochituate. These were built after the Middlesex Canal ceased to operate in practice, though its charter remained legal until 1856. These waters were held in the winter and during times of freshet, and then released exactly when the manufactories needed them in the late summer. As a result, in July and August they "pour into our valley every day for these months, upward of seventeen millions of gallons," whereas the natural flow from the lake was only "five millions daily." These "deluging powers" or "water avalanches" kept the lower reaches of the river full of water and made the meadows wet and unusable.[7]

The haymakers—trapped between the metaphorical alligator in the water and the tiger on the shore—were getting less hay, and what they did get was of inferior quality, and only a few cranberries during drought years.

Their land had greatly depreciated. "Sixty-five years ago, the Meadows were perfectly accessible to the heaviest teams, up and down the river, to its brink, . . . without the slightest difficulty from slumping. . . . But from about the year 1804, when the Canal Proprietors had made two additions to the height of their dam, and had opened the Canal for travel and transportation, the Meadows became so soft as to be impossible for teams, except in times of extreme dryness, and then only for light ones, and in particular parts. Since the last addition to the dam, thirty years ago, these lands have been, with slight exceptions, inaccessible."[8]

Another of the meadow owners' advocates, Judge Abbott, then asked: "And now that the Compensating reservoirs are no longer needed, what does the Cochituate Water Board do? Instead of relieving the meadowland farmers below, they have gone and sold the Marlborough Reservoir to a mill-owner, for a trifling consideration . . . [and] . . . [t]he Hopkinton establishment they have leased for ten years, at a few hundreds." Abbott then explained the ethical rationale for bringing the story out at this time: they hoped it "may prove instructive, and we gladly embrace an opportunity to bring to the bosom of a great and powerful Commonwealth, a story of strange complications, not to say chicaneries, resulting in a practical and persistent denial of justice to an inoffensive and loyal community, flagrantly wronged and outraged, through two entire generations."[9]

The final part of the petition was the legal part: "We believe that the meshes of the Middlesex Canal Act were woven by the subtle fingers of lobby legislation." After narrating a compelling tale of corruption, they identified five specific legal complaints against the canal legislation. First and most insightful was the issue of complexity, aka wildness: "No man, nor set of men, can for[e]see the extent, nor all the incidents, nor, in all respect, the *nature* of the damage which may accrue." Correctly they pointed out that there was no way to know how the consequences of raising a dam would ramify through the system. A second concern was that the act of incorporation for the canal allowed only one year for a plaintiff to prosecute for damages, even though the damage accrued annually. This one-year statute of limitations made no sense in a climate where the variability in annual rainfall was greater than the average annual amount. In such a setting, it would take several years to notice a real long-term change. Third, the appropriation of the water of the Concord River by the Canal Company had been a covert decision made before the farmers had a chance to object. It was never "openly proposed" in such a way that the meadow owners could legally respond. Rather, the river's water was "stealthily usurped." Fourth, no tribunal had been created by the Canal Act that was "competent to determine what should be the height of the Billerica Dam; simply because there is not a magistrate, nor a court, from the lowest of the highest, in the State, that has the slightest power of the subject, any more than the puniest child who breathes the malaria which it generates." That is, the dam height was completely arbitrary, and its indirect impacts beyond the immediate flowage of the dam pool were unknown. Fifth, "though the power to" claim the water and build a dam "lay within the Canal's authority, they did not give back the rights to the water to the meadowlands when the water was no longer needed, but sold it to a private miller for probably less than a quarter of what it was worth in power, rather than give it back to us." In short, the river was being treated as a commodity taken from the farmers by the state and sold to private industry for a profit.[10]

After presenting their case, the petitioners requested four things: first, that the legislature authorize the appointment of a commission to investigate the matter, take testimony, and examine the situation in the field; second, that the one-year limit of time to prosecute claims be extended, so that the meadow owners could sue; third, that the canal proprietors' charter be declared void and the corporation dissolved; and fourth, that the time limit be extended for assessing damages against the city of Boston for "an Act to supply the City of Boston with Pure Water."[11]

Boating for Pay

As the petition was working its way through the legislature in early 1859, Henry Thoreau was enjoying boating on a dramatic spring freshet. On March 7, out of the blue, he had a remarkable insight that would forever change the way he did things and would help motivate his later project. He realized that "the cause and the effect" of many natural phenomena "are equally evanescent and intangible, and the former must be investigated in the same spirit and with the same reverence with which the latter is perceived." More generally, he came to see that the ways and means of "the economy of Nature" could be understood, and that this understanding could enhance his appreciation of nature.[12]

Probing causes beneath appearances is the essence of what historians do. It's the essence of what scientists do. It would become the essence of what Thoreau would do for the rest of the summer. The effects of flowage were self-evident: the meadows were wetter than previously, the bridges were constricting the flow, the bars were greater obstacles, and the meadows were more pockmarked by ice rafting. But what were the underlying causes of these appearances? That summer he would find out by becoming a silent third party in the flowage controversy. In contrast, the opposing attorneys for the petitioners and the respondents were doing what they were trained to do: shape the appearance of things in such a way as to support a priori beliefs, and then cherry-pick the causes (explanations) that support those beliefs.

Working on his own, Thoreau inaugurated a truly scientific investigation of the largest, most powerful, and wildest thing in his life, the Concord River, by which I mean the entire watershed above its natural outlet at the Fordway. Learning its habits would require what modern scientists call "inverse reasoning." This procedure answers the question "Given this, what happened?" Getting an answer requires that the "given this" part of the question be extensively documented. Hence Thoreau's mapping, measuring, and monitoring of the Musketaquid valley required weeks of commitment to data collection and report writing. The second side of the inverse reasoning question, the "what happened," required genius-level inductions and explanations based on the observations.

Ralph Waldo Emerson was an eyewitness to Thoreau's practice that summer: "Every fact which occurs in the bed, on the banks or in the air over it . . . [was] known to him. . . . He liked to speak of the manners of the river, as itself a lawful creature, yet with exactness, and always to an observed fact." In short, Thoreau used science to animate the river in his mind, to find the laws by which this creature lived its life. Though constrained by bridges, causeways, and canals, it was no more tamed than a caged wild animal, because every agricultural and engineering intervention created a cascade of effects that ramified up or down the valley in unpredictable ways and without human oversight.[13]

ON MARCH 15, 1859, the Committee of Agriculture of the Massachusetts House of Representatives, chaired by John S. Eldridge, reported that it had considered the petitions requesting the removal of nuisances and unauthorized encroachments and made the following decision: "*Resolved,* That a joint committee, consisting of two members on the part of the Senate, and three members of the House of Representatives, be appointed to investigate the flowage of lands lying in the towns of Concord, Wayland, Sudbury, Bedford, and Carlisle, and to recommend such changes and improvements in said flowage as they may deem requisite for the public good and the owners of said lands." By the end of the month, the resolution had been "engrossed," and it was sent for concurrence to the House and Senate on March 24 and 28, respectively. On April 1, the General Court (legislature) of the Commonwealth passed this first of four separate acts regarding the flowage controversy.[14]

On March 17, coincident to the day with the appointment of the Joint Committee, Henry began a program of systematic stage measurements near the river's triple point. His original datum was the base of the slanting iron truss of the railroad bridge, located four feet from its eastern end. At the time, the water was rising. On March 17 he caught the peak of the spring freshet, recording it to be five inches from the highest part of the truss. When making these measurements, he saw and sketched the water surface "heaped up in the middle" between each of the sets of piers supporting Flint's Bridge. He'd been watching this sort of thing since the fall of 1853, when he noticed that each bridge acted like a dam during strong flows because the abutments confined the flow and the piers partially blocked it. The result was a river surface elevation on the upstream side of each bridge that was at least several inches higher than on the downstream side. This effect was exaggerated during high flows, especially when breakup ice was present during the spring freshet.[15]

Thoreau realized that river stage at any point was much more complex than a dipstick measurement. Rather, it was a function not only of the water budget but also of the strength of the current, the degree of constriction, and the presence or absence of objects such as ice floes and meadow tufts. Stated conversely, each bridge backed up the river in its own way. Stated mathematically, if one knew the input variables, one could calculate the output effect for each bridge, all the way up to the high side of the highest bridge. This is exactly what Thoreau was contracted to do three months later, which suggests that the bridge work was his idea.[16]

Observing the spring freshet of 1859 led him to a separate radical theory about how the bridges and causeways were permanently changing the meadows. Each bridge interfered with the transport of ice-rafted clumps or tufts of meadow. Without bridges, clumps were rafted straight downstream and/or directed by the wind. But with bridges, clumps were either stopped by the causeways or concentrated by the raking action of bridge piers: "Some meadows are now saved by the cause ways & bridges & willow rows." Over time, what had originally been a gently sloped alluvial valley before bridge construction was becoming a series of steps, with flatter, more sluggish treads between bridges and zones of steeper, faster flow near them. This correct theory—which he developed before he was hired—became part of the text of the 1860 flowage report. It's quite likely that Thoreau shared it, perhaps with Simon Brown, the president of the committee, whose land Thoreau visited that spring while searching for arrowheads.[17]

The Joint Committee appointments were made by the legislature on April 6. The chair, Mr. Wrightington of Fall River, kept a beautifully inscribed notebook of the proceedings, now available for perusal at the Concord Free Public Library. On April 8, only two days after the joint legislative committee was appointed, Henry ranted in his journal about the process of appointing legislative committees. Coincidence? Possibly. Though his rant uses songbirds as an example, the timing suggests it was an allegory about the political machinations of committees in general.[18]

Family and financial considerations that spring may have helped motivate Thoreau to push his ideas forward to the River Meadow Association in the hope of obtaining paid work. His father had died on February 3, making Thoreau the legal head of household, and requiring him to post a $10,000 bond as executor of his father's estate. He also bought a self-help book, *The Businessman's Assistant,* to help the family manufacturing business. On April 3 Thoreau complained in his journal that his lecturing and writing were not bringing in much revenue. On the eighteenth he hired himself out to survey someone's land for development, and on the twenty-first he agreed to work for a short period as a manual laborer, planting trees. From May 3 to 17 he performed a batch of land surveys that overlapped with the May 11 opening of Joint Committee hearings in Boston. All these jobs suggest that Thoreau was then being motivated by financial considerations.[19]

ON MAY 23 THE COMMITTEE HEARINGS MOVED from Boston to the Concord courthouse. For the first time, the opposing parties came face-to-face. The petitioners had representatives from six meadowland towns; the respondents had representatives from two factories and the rocky Town of Billerica, where many factory workers lived. On this first day in Concord, the Joint Committee had only three of its five members present. Their first order of business was to schedule a pair of day trips, upriver on the twenty-fourth and downriver on the twenty-fifth.[20]

On the upriver trip, they found conditions to be lakelike: "a vast expanse, was either completely covered with water, or rendered soft and yielding to the step, making a passage across it decidedly unpleasant . . . meadows, so called, were one vast sheet of water." By the second day, all five committee members had arrived, joined by the Hon. Charles R. Train, representing the City of Boston. Heading downriver on the twenty-fifth, they viewed "vast tracts of meadow-land completely covered with water, and rendering the lands farther back from the River, soft and spongy." This inundation ended at the Fordway, a "natural ford, which the Remonstrants [respondents] say more effectually impedes the progress of the waters than does the Dam." This is true, but what they did not realize at the time was that the dam pool greatly decreased the efficiency of the outlet by reducing its downstream gradient. "At the dam, the water seemed to be about one foot higher than the flash boards over which it flowed." This indicates strong flow in the river combined with the factory's policy of holding back as much water as possible.[21]

Henry Thoreau got hired as a consultant on June 4. Specifically, a subcommittee of the petitioners led by town leaders—Simon Brown and Samuel H. Rhodes of Concord, David Heard of Wayland, and John Simonds of Bedford—signed a letter to Henry Thoreau asking him to carry out several tasks related to their legal case. They represented a group called the River Meadow Association, the name Henry listed in his survey notebook and in a letter dated July 8. The full membership of the association, listed in a questionnaire sent out to individual property owners later that summer, had voted to approve his selection. This action is indicated by one of Thoreau's private letters, suggesting he was present for the vote.[22]

Hiring Thoreau for a case of wetlands flowage was not without precedent. Six years earlier William Benjamin, of Lincoln, had raised his dam to strengthen the power of his mill, flooding upstream landowners. One of those whose crops were flooded, Leonard Spaulding, sued for damages in the Middlesex Court of Common Pleas in February 1853. Before the case came to trial, Spaulding hired Henry to measure the height of the dam and thus the flowage. To his surprise, Thoreau was served with a summons compelling him to appear as an expert witness, and he spent three days in court. But Spaulding lost the case.[23]

The June 4 letter to Henry was written in pen on one side of a page torn from a notebook. The script is beautiful, almost calligraphic. Unfortunately, all that remains of that letter is the last page of an unknown total. In full, that reads:

> Notice that the width of the bridges from Wayland to the dam at Billerica be measured between the abutments. The number and size of the wooden piers and of the stone piers. The depth of water at each bridge. To learn, if possible the time of erection of each bridge, and if any abutments have been extended since the building of any bridge, & when. And to ascertain the width of the river at its narrowest place, at the falls, and the capacity of the sluiceway leading toward ~~Lowell~~ merrimac. And if any bridges have been discontinued.

They wanted him to perform three specific tasks: to inventory, date, and measure specific attributes of existing and historic bridges; to measure the width of the river at its narrowest place, here called the "falls"; and to estimate the capacity of the sluiceway of the Concord River leading toward the Merrimack River, which lies below the Billerica dam. These tasks involved civil engineering and history, rather than surveying and science. Thoreau summarized his bridge work in an elaborate "Table of Statistics of Bridges," which went far beyond what he had been asked to do. For the two specific measurements—the narrowest place and the sluiceway capacity—and the bridge history, he entered results directly into his private journal. There we learn that the narrowest place was not at the falls, as the letter indicates, but in the upper Sudbury.

Thoreau may have thrown away the rest of the appointment letter, keeping the last page because he had written on its back (in his hand in pencil) a list of three items to procure and the names of eight individuals in four adjacent towns to contact. All would later testify in court, and most were mentioned in Thoreau's journal entries.[24]

The only other concrete evidence for Thoreau's employment is a cryptic entry in his survey notebook, which at this stage was little more than a calendar inventory of the projects he was hired to carry out: "River meadow association June (21) 22 - 3 - 4 & / 1 day more—." From the pencil and spacing, it appears that the phrase "& 1 day more" was added later. This suggests an agreement for four days' work, with a fifth added, perhaps the limit of the association's budget at the time. Nevertheless, Thoreau would go on to work nearly full-time for two more months without pay.[25]

Henry may have received his June 4 appointment letter on June 11, based on the sequence of entries in his survey notebook. On its own line in pencil, it follows "Clark June 10—59 v. plan" in ink. It precedes "Cyrus Stow [illegible] for July 11." We know he was working on June 16 because that's when he mentions in his journal that he will soon be measuring bridges. June 18 is the postmark date for an envelope on which he was creating a table of data associated with the bridge statistics. On the reverse side of that letter he began compiling a list of treatises on hydraulics: "Robison on Rivers" and " 'Lowell Hydraulic Experiments' by James B." On a separate piece of notebook paper, these references are relisted more completely and crossed off, likely indicating he later located and used them.[26]

Based on these four known records—the appointment letter, Thoreau's survey notebook, his correspondence, and his journal—Thoreau was not "hired by the River Meadow Association to help survey the river in preparing their case against the dam," as Patrick Chura claims in his biography of Thoreau's surveying practice. Thoreau did no surveying. That had been done by Loammi Baldwin in 1811 and then by B. P. Perham in 1833 for what became known as "Baldwin's second map." And though Thoreau made hundreds of quantitative measurements, especially depth soundings, there's no evidence that he was ever asked to do so by his client. Rather, these activities were part of his parallel, apparently unpaid river project that seems to have been kept out of the official investigation. The Concord Free Public Library, which holds these materials, the editors of the Princeton edition of the *Journal,* and Thoreau scholars consistently refer to this work as the "Concord River survey" or "river survey." This is a misnomer. Thoreau never used this phrase to describe his work. The phrase "river project" is both more accurate and inclusive because Thoreau collected lots of quantitative information—velocities, temperatures, fluxes, and classifications—having nothing to do with spatial measurement, and the only maps he produced were based on much older surveys done by others.[27]

During the summer of 1859, Thoreau worked nearly full-time for eight weeks. And in 1860 he worked nearly full-time for another two weeks and part-time for months. The totality of his project spans eighteen months, from

March 17, 1859, to September 27, 1860. To my knowledge, Thoreau's river project is the most wide-ranging scientific (i.e., theoretical) investigation of any American river prior to 1877, when Grove Karl Gilbert, a charter member of the U.S. Geological Survey, reported on the streams of the Henry Mountains in Utah. This claim excludes the equally sophisticated engineering studies of America's larger rivers because the interest there didn't involve natural science. Thoreau's study is a pioneering examination of a disrupted river system that predates, by half a century, Gilbert's early twentieth-century study of rivers impacted by gold rush mining in California.[28]

Transition

Henry's transition to river work began on June 16. That's when his normal routine of botanical inventory abruptly shifts to a systematic transect across the Great Meadow. The result was a flora based on distance from the natural levee of the Concord River and the elevation above standing water. He repeated this transect on July 7 suggesting he was monitoring botanical change through time. Among the plants, he lists six different kinds of sedges alone. I interpret this task as the first of many he carried out for himself that summer. His flora matches the nineteenth-century summary provided by historian Brian Donahue from other sources, with meadows "dominated by cordgrass (*Spartina pectinata*), reed canary grass (*Phalaris arundinacea*), fowl meadow grass (*Poa palustris* and *Glyceria striata*), blue joint (*Calamagrostis canadensis*), red top (*Agrostis alba*), and a number of other grasses and sedges."[29]

Much of the quantitative data in Thoreau's journal for this interval involved keeping track of the water level, or river stage. Spring floods came consistently every year, though their peak stage varied considerably. Summer and autumn floods were unpredictable because they involved major subtropical storms, some of which were wet hurricanes. As early as June 1851, Henry found it hard to "believe that there is so great a difference between one year and another as my journal shows." The following year showed even greater extremes. Here was an inland river whose elevation in Concord routinely fluctuated nearly nine vertical feet. "The river is eight and one twelfth feet below the top of the truss. Add eight and a half inches for its greatest height this year, and you have eight feet nine and a half inches for the difference. . . . That is, those are the limits of our river's expansibility; so much it may swell.

Of course, the water now in it is but a small fraction of that which it contains in the highest freshets."[30]

When beginning his first systematic study of river stage, Henry found it frustrating that such an important parameter had not been kept track of as a matter of public record, like a census of population or a precipitation record. Earlier he had decided that Concord "should provide a stone monument to be placed in the river, so as to be surrounded by water at its lowest stage, and a dozen feet high, so as to rise above it at its highest stage; on this feet and inches to be permanently marked; and it be made some one's duty to record each high or low stage of the water. Now, when we have a remarkable freshet, we cannot tell surely whether it is higher than the one thirty or sixty years ago or not. It would be not merely interesting but practically valuable, to know this."[31]

The closest thing Musketaquid had to a local stage gauge was a large boulder known as Saddle Rock, located at Heard's Bridge at the southern end of Sudbury Meadow. This stage record had been kept for many years by Richard Heard, who had been involved in the 1834 lawsuit and who was the father of one of the 1859 ringleaders.

In lieu of a public data point in Concord, Henry created his own near his boat place on or before June 20, 1859. This unequivocally marks the onset of his river project. He referenced this datum to three backup sites, as well as to "summer level," the stage marking the threshold between flood runoff and groundwater drainage. Lost in these details is Thoreau's scientific method. He understood that for his bridge measurements and soundings to have meaning, all his depths and widths over a distance of twenty-five river miles must be tied to a common datum. He was fully aware of the considerable errors introduced by wind shear, which pushed water upstream (and prevented driftwood from descending), and by the harmonic sloshing of water within open reaches, called seiche waves. He kept his stage measurements on scraps of paper and in notebook entries before transferring them to his journal as raw data, crude tables, and numbers in text. What had been a literary manuscript for years was suddenly becoming a poorly written mash-up of run-on sentences and fractional numbers accurate to the nearest sixty-fourth of an inch. He consistently avoided decimal fractions.[32]

Thoreau's early botanical work and preparation for the bridge survey in mid-June was interrupted by "rain—esp—heavy rain raising the river in the night of the 17th." Though he attempted to get started on the eighteenth, the rain was still so heavy, it forced him to take shelter all day under bridges and his overturned boat. In response to this rain, the river at Saddle Rock rose nearly 30 inches, to a peak of 42 inches on June 22. After this flood crest, the upper Sudbury began to drop at an accelerating rate as the slug of water migrated downstream toward Concord. Thoreau used his local stage monitoring to show that this mass from the Sudbury arrived in Concord well after the flood wave of the Assabet. His local data also showed that the flood crest at Concord came two days after its crest in Wayland, and that its decline had been prolonged by a backup of water at the Fordway outlet.[33]

Sometime before June 18 Henry had Baldwin's 1834 survey map in hand, because he was already taking notes from it. That date is the postmark on a postcard addressed to "Mr. Henry D. Thoreaux," and paid for with a three-cent stamp of George Washington. On the backside of the card are measurements and calculations regarding the distances between bridges "by Plan and by Me." This work translated the straight-line distances of Baldwin's surveys to the natural curves of the river.

Bridge Statistics

A critical issue for the meadowlands controversy was the extent to which bridges raised the water level on their upstream sides. Thoreau knew that hydraulic damming at bridges was occurring. For typical, ice-free, summer conditions, a bridge might raise the water "several inches" on the upstream side, thereby creating a "slight rapid" below each bridge. The problem for the petitioners was that these effects are cumulative. Thoreau knew that three inches for each bridge multiplied by twenty bridges yielded sixty total inches of rise for the most upstream meadow. This calculation was consistent with the fact that the greatest problems with summer drainage lay farthest upstream. Indeed, this estimate was close to the total surveyed drop in river level, or the fall.[34]

On June 22 Henry finally got the chance to begin his official bridge survey work. For that day's journal entry he wrote: "Paddle up the River to Lees measuring the bridges. The sun coming out at intervals today—after a long rainy & cloudy spell—in which the weeds have grown much." The next day he traveled to Wayland by road to work with David Heard, the man who paid him for his work for the association. That same day Thoreau returned to Concord and paddled all the way down to the Falls in Billerica, assisted by Ellery Channing. Channing recalled this work in his biography: "From Billerica Falls to Saxonville ox-bow, thirty miles or more, he sounded the deeps and shallows of the Concord River, and put down in his tablets that he had such a feeling." I interpret the phrase "had such a feeling" to indicate Thoreau's high and low moods about the project. By the twenty-fourth Henry was done with his bridge work and had already nearly used up his job budget. As a (presumably unpaid) adjunct to this work, he apparently measured "all the swift & the shallow places" between the bridges.[35]

Henry's other consulting task was to gather information about bridge history. His journal for June 24 is a repository of oral history that later showed up in court testimony. For example, "John Hosmer tells me that he remembers Maj. Hosmer's testifying that the south bridge was carried up stream—before the court at the beginning of the controversy." Having the South Bridge wash upstream indicates a powerful flow reversal caused by the Assabet, perhaps in conjunction with strong northerly winds and river ice. Also in his journal are the records of personal interviews he conducted about the general conditions of bridges in the early decades of the nineteenth century, mostly in Wayland and Sudbury. For example, the water below Heard's bridge at the Wayland/Sudbury line was so low in the early nineteenth century that boys could wade it easily, whereas at the time of Thoreau's investigation the water was always much deeper. The water level of the local meadows "was much lower then in summer than now." One of the very best meadows had been abandoned. The dam for the Sudbury Canal, now wet, used to be dry. He was also told that the current of the upper Musketaquid was now more sluggish than earlier, owing to Assabet back-flowage as far as the Sudbury meadow.[36]

In addition to personal interviews, Thoreau sent off letters soliciting information about bridge history from local experts. The latest of the return letters is from Jon Hill, of Billerica, dated July 25, 1859, which is three full weeks after Thoreau turned his attention away from what he was hired to do and began focusing on other river project tasks. The most interesting letter was a July 1 letter from David Heard, of Wayland. In the salutation, Thoreau is addressed as "Thorow," the phonetic punctuation of his name. He was also given the honorary title "Esq.," which is generally reserved for professional (often legal) work. Both letters indicate that Thoreau had left his employment with the River Meadows Association on good terms. I've found no invoice or accounting of how much Henry was paid, suggesting this was a "need-to-know" cash deal on the honor system.[37]

AT SOME POINT THOREAU SUMMARIZED his bridge work in a magnificent table with the cumbersome title "Statistics of Bridges over Concord River between Heard's Bridge and Billerica Dam, obtained June 22d 23d, & 24th 1859; the level of the water at Concord, in the meanwhile, not having rained one inch from about 3 feet above summer level." In it he provides sixteen columns of information for twenty-one bridges, four of which had been discontinued and thus were not measured. The only column of hydraulic relevance used for his later analysis of river behavior gives the peak stage for that year's March freshet at each of seventeen bridges. He found that its upper surface was essentially a horizontal plane, making Musketaquid less a stream than a large, ribbon-shaped lake extending more than twenty miles above its natural outlet at the Fordway. Superficially, it resembled a narrow version of the Florida Everglades, being grassy and infinitesimally inclined. From his field measurements on bridge abutments, Henry found the surface of the water during the freshet to be 4.39 feet above the stable summer level. This compares almost exactly with the total fall in the river of 4.32 feet for the 1811 survey. His journal entry for June 22 correctly attributes this lakelike condition to hydraulic damming at the Fordway, which is similar to what happens when the drain for a lavatory sink is too small for its wide-open faucet. It backs up.[38]

Henry's loose notes for the river project in the archive of the Concord Free Public Library contain three previous drafts for "Statistics of Bridges." In the penultimate draft he had columns for "character of current" and "character of bottom," neither of which survived the last revision. These simplifications suggest some back-and-forth between client and consultant before the final decision was made to exclude Henry's work from the legal proceedings.

Never does the graphic quality of his master table of bridge statistics rise to the visual standard of a courtroom document. There is no calligraphy and no clean title, many numerical values are crossed out and written over, several ink blotches were written around, and many words are indecipherable. Units of measurement are not given. And, most important, the table is meaningless unless cross-referenced to his explanations in his private journal. If Thoreau submitted a final copy or a report to his clients, it has not yet been discovered.[39]

There are only three explanations that make sense for why Thoreau apparently never finished the bridge task he was hired to do. First, he may have actually completed a finished bridge document and report, but the final copy either disappeared or was intentionally destroyed. Second, he may have had a falling-out with the committee that hired him and was no longer willing to work to serve their purposes. Third, and most likely, he learned something that would have undermined the case for his clients, and both parties agreed to keep it quiet. For example, his flood profile for the March freshet showed that the Fordway, rather than the Billerica dam, was the main cause of flooding. This would have strengthened the case for the opposition. Additionally, his bridge statistics, if publicly available, would have given the respondents what they needed to calculate the grand sum of hydraulic impacts for all twenty bridges. Finally, keeping him off the witness stand prevented the possibility of him making a mockery of both sides.

9 | Mapmaker

of paid work on the bridge project. There's no entry for June 25, perhaps because the entire day was spent compiling his draft table of statistics. After that, Thoreau seems to have returned to his normal sojourning activities: taking a boat trip up the Assabet on the twenty-sixth, making two trips to Walden Pond on the twenty-seventh and twenty-ninth, and resuming his natural history work of botanical phenology.

By July 4, however, he was back at work—apparently for himself—on what would turn out to be a vastly expanded river project. Between Independence Day and the eighth, he worked nearly full-time on a mapping-measuring program that gave rise to his scroll map. He characterized the size, shape, and character of river channels along the full length of Musketaquid between Saxonville and North Billerica. He also gave specific attention to an inventory of conspicuous deeps and shallows, which eventually led him to classify seven distinct river reaches. Initially he recorded anecdotal observations and measurements in his journal in the order they were encountered. A representative quote is: "P.M.—To FH Pond, measuring depth of river . . . The deepest place I find in the river to-day is off Bittern Cliff, answering to the bold shore. There is an uninterrupted deep and wide reach of the river from Fair Haven Pond to Nut Meadow Brook." By July 7, however, he decided he needed a base map on which he could compile all of his observations and which might reveal clues to underlying causes. While looking at "Baldwin's second map," which was created for the 1834 trial, he noticed that the Musketaquid was so nicely aligned in one direction that "it can be plotted by the scale of 1000 feet to an inch on a sheet of paper 7 feet 1¼ inches long by 11 inches wide."[1]

He followed through with this inspiration by making a map on tracing paper and gluing it to a spliced roll of surveyor's cloth, probably to make it durable enough for field use. Its many smudges, wrinkles, and frayed edges are consistent with this interpretation. Left to cure for 155 years, the resulting scroll map is now discolored to a mottled light reddish brown, and has stiff-ened into a tight scroll. The Concord Free Public Library's inventory de-scribes this map unrolled as 91 inches long (7 feet 6½ inches) by 15 inches wide, which is five inches longer and four inches wider than needed to map the channels of Musketaquid. This extra width gave him the room to add the vertical profiles of water surface elevation from two previous survey maps. When everything was plotted and notes were displayed, this scroll map be-came the graphical key to Thoreau's whole river project. Without it, there is no spatial anchor for locations and elevations, and no visual image from which to interpret the channel patterns and sequence of reaches.[2]

On the tracing paper overlay are a series of crosscutting lines for different stages of map production. Oldest is a light gray tracing made from Baldwin's 1834 map, which was surveyed by B. P. Perham in 1833. Its complete title was carefully replicated in black ink in calligraphic style on the right-hand side of the map:

PLAN
of
CONCORD RIVER
from East Sudbury to Billerica Mills,
22.15 Miles.
To be used on a trial in the S.I. court,
Sudbury & East Sudbury Meadow Corporation
vs.
Middlesex Canal.
Taken by agreement of Parties,
By
L. Baldwin, Civil Engineer.
Surveyed & Drawn by B. P. Perham.
May 1834.

The civil engineer in charge was Colonel Loammi Baldwin. After grad-uating from Harvard at the age of twenty and helping to design the Bunker Hill Monument, he became an expert on tapping "the power of enlisting Na-ture as an aid and of turning her forces into profitable channels." His integrity and expertise were so respected that both sides of the 1834 court case trusted him as a resource. Note that this 1834 prequel to the 1859 flowage controversy

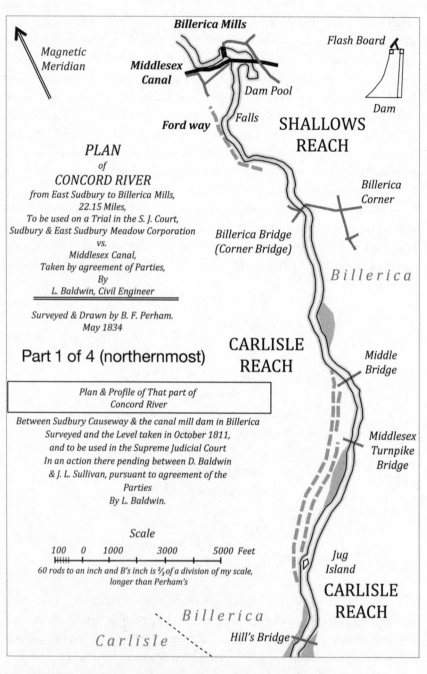

Billerica Mills

Magnetic Meridian

Middlesex Canal

Ford way *Falls* **Dam Pool**

Flash Board

Dam

SHALLOWS REACH

PLAN

of

CONCORD RIVER

from East Sudbury to Billerica Mills,
22.15 Miles,
To be used on a Trial in the S. J. Court,
Sudbury & East Sudbury Meadow Corporation
vs.
Middlesex Canal,
Taken by agreement of Parties,
By
L. Baldwin, Civil Engineer

Surveyed & Drawn by B. F. Perham.
May 1834

Part 1 of 4 (northernmost)

Plan & Profile of That part of Concord River

Between Sudbury Causeway & the canal mill dam in Billerica
Surveyed and the Level taken in October 1811,
and to be used in the Supreme Judicial Court
In an action there pending between D. Baldwin
& J. L. Sullivan, pursuant to agreement of the
Parties
By L. Baldwin.

Billerica Corner

Billerica Bridge (Corner Bridge)

B i l l e r i c a

CARLISLE REACH

Middle Bridge

Middlesex Turnpike Bridge

Scale

100   0   1000   3000   5000 *Feet*

60 rods to an inch and B's inch is ⅔ of a division of my scale,
longer than Perham's

B i l l e r i c a

C a r l i s l e

Jug Island

CARLISLE REACH

Hill's Bridge

FIGURE 20A. Simplified scroll map, part 1 of 4. Northernmost portion of Thoreau's
scroll map showing Shallows and Carlisle reaches. See text for details.

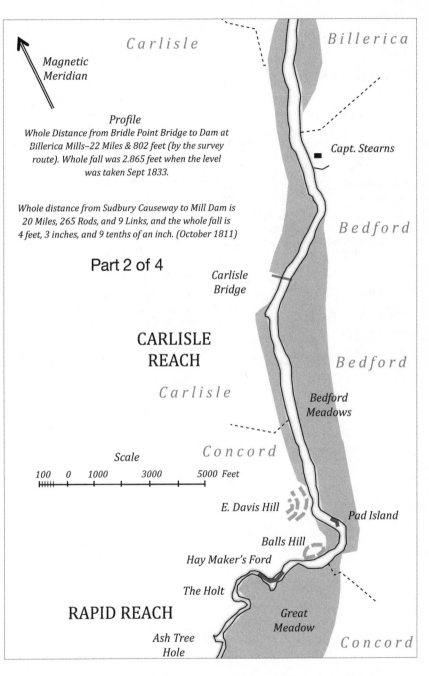

Carlisle

Billerica

Magnetic
Meridian

Capt. Stearns

Profile
Whole Distance from Bridle Point Bridge to Dam at
Billerica Mills–22 Miles & 802 feet (by the survey
route). Whole fall was 2.865 feet when the level
was taken Sept 1833.

Bedford

Whole distance from Sudbury Causeway to Mill Dam is
20 Miles, 265 Rods, and 9 Links, and the whole fall is
4 feet, 3 inches, and 9 tenths of an inch. (October 1811)

Part 2 of 4

Carlisle
Bridge

CARLISLE
REACH

Bedford

Carlisle

Bedford
Meadows

Concord

Scale

| 100 | 0 | 1000 | 3000 | 5000 | Feet |

E. Davis Hill

Pad Island

Balls Hill

Hay Maker's Ford

The Holt

Great
Meadow

RAPID REACH

Concord

Ash Tree
Hole

FIGURE 20B. Simplified scroll map, part 2 of 4. Portion of Thoreau's scroll map showing Carlisle and Rapid reaches. See text for details.

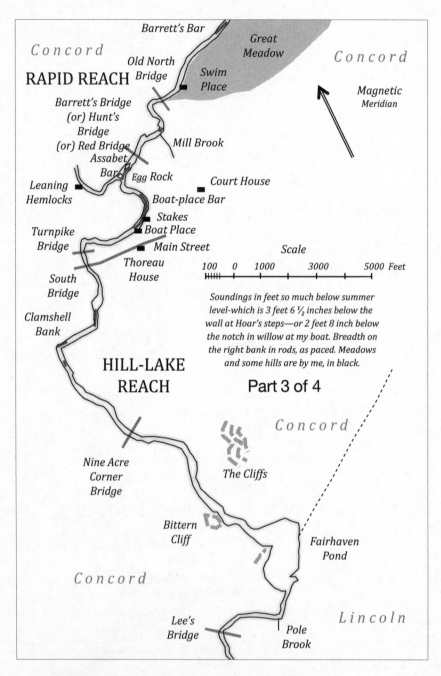

FIGURE 20C. Simplified scroll map, part 3 of 4. Portion of Thoreau's scroll map showing Rapid and Hill-Lake reaches. See text for details.

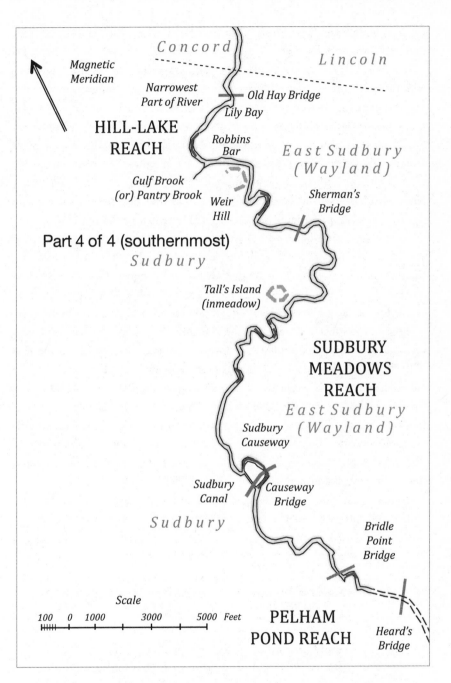

FIGURE 20D. Simplified scroll map, part 4 of 4. Southernmost portion of Thoreau's scroll map showing Hill-Lake, Sudbury Meadows, and Pelham Pond reaches. See text for details.

was launched from "Sudbury & East Sudbury," where the wettest postcolonial meadows were located.[3]

Thoreau's use of Baldwin's title, rather than one of his own, together with the careful rendering and placement of the title, town names, and key geographic features, suggests that his original plan was to create his own version of Baldwin's map, possibly for use as a public document, perhaps even for court testimony. It's plausible that he was hired separately for a few days to make such a map, although this is pure conjecture. What happened instead, maybe for reasons similar to those involving the "Statistics of Bridges" table, is that the River Meadow Association decided to use Baldwin's original map as the sole illustration to support their case, and leave Thoreau's work out of it.[4]

Using the same black ink as for the title, he drew a 5,000-foot-long graphic scale that used the personal pronoun "my," proving that that the map was, by that time, being used for private purposes comparable to that of his private journal. Indeed, map and journal reference each other. On his extra-wide copy, Thoreau then drew a series of stepped lines that tracked the progressive rise in the surface elevation of the Concord River between bridges, based on the 1833 topographic profile. The "whole distance" of the profile is the sum of the horizontal line segments, in this case 22 miles and 802 feet. The "whole fall" is the sum of the vertical line segments, from the dam upward. Dividing the fall by the distance gave the average slope of the water surface, which in this case is 2.865 feet of vertical drop over a distance of 116,902 feet, or a dimensionless ratio of 0.00002, or about two parts in 10,000. This was a truly negligible slope for a midsized river, comparable to the lower Amazon. Except in narrows and shallows, Musketaquid was effectively a lake.

Using a different ink that aged to a rich brown color and bled into the paper, Henry transferred a second set of features from a previous survey map:

Plan & Profile of that part of
Concord River
between Sudbury Causeway & the Canal Mill Dam in Billerica,
Surveyed & the Level taken in
October 1811,
and to be used in the Supreme Judicial Court,
in an action there pending between D. Baldwin
& J. L. Sullivan, pursuant to agreement of the parties.
By L. Baldwin.

This map was made by the same Loammi Baldwin when he was twenty-two years younger and working for his brother, D. Baldwin, in a lawsuit against the state-chartered Middlesex Canal, rather than against private factories. The most important thing Thoreau transferred was the river profile. The sum of links from the 1811 survey showed a greater fall of 4.325 feet over a similar distance of 109,978 feet, for an overall slope of 0.00011. Though still very low, this slope is five times steeper than that of the 1834 map. Was the difference due to an error? Or had the river gradient been lowered one and a half feet by the raising of the dam? To solve this discrepancy, the river was resurveyed even more accurately in 1859 by John Avery, who was working for the respondents. His resulting river profile split the difference between the 1811 and 1833 surveys, finding a fall of 3.04 feet over a distance of 22 miles, which was ten inches greater than the 1833 survey.[5]

To his scroll map, Henry transferred three other items from Baldwin's 1811 map, as shown by the brown ink: an accurate cross-section of the dam; a sketch of the Floating Bridge across the dam pool; and, most important, a set of descriptions of the bed conditions for well-known bars in the channel, mainly in Concord. This proves that the sediment bars were being inventoried and investigated during the 1811 court proceedings as potential causes of chronic upstream wetness. It is significant that the long, shallow bar near the mouth of the Sudbury River (where Thoreau kept his boat) is completely missing from the 1811 inventory of channel obstructions, even though Thoreau would later map it in 1859 as the longest, most conspicuous bar in the whole river. Critically, this was the bar responsible for creating lakelike conditions in Musketaquid as far south as Sudbury meadows. It seems not to have been there in 1811.[6]

The span of time between these two surveys taken twenty-two years apart (1811 and 1833) gave Henry a chance to compare before-and-after river gradients for eighteen reaches between twenty bridges. Fortuitously, this time interval captured two critical events in river history that Thoreau would soon be investigating. In 1827 or shortly thereafter, the Union Turnpike Corporation built a one-arch stone bridge over the lower Assabet River. Its design was too narrow, it was mislocated on a bend, and it was misaligned relative to the streamflow, causing major erosion of the bank, deep scour, and the delivery of coarse sediment down the main stem toward the T-junction. One year later, in 1828, the capstone sills were laid for the newer, higher, and stronger stone-faced Billerica dam, and the water raised an additional fifteen inches with flashboards.[7]

The perturbation caused by these two engineered structures—the 1827 bridge and the 1828 dam—almost certainly account for the dramatic change in the before-and-after river profiles that Thoreau recorded immediately above and below the mouth of the Assabet River. To the north (downriver), the slope increased dramatically. To the south (upriver), it decreased dramatically. Changes further away in both directions were negligible. Seeing these differences likely prompted Thoreau to look for a common cause, which he later realized was a pulse of coarse sediment pouring down the Assabet (caused by the 1827 bridge) that was deposited as a pair of flanking gravel bars whose tops were higher than before because they were deposited in higher floods associated with the higher 1828 Billerica dam. In the upstream direction, deposition of a higher boat-place bar diminished the preexisting uniform slope. In the downstream direction, deposition of a higher Assabet bar increased it.[8]

ON HENRY'S SCROLL MAP, there are faint marks in light gray pencil left after erasure that show Baldwin's generalized 1834 river channel. In a recent analysis, Daegan Miller writes that Thoreau's river channel was "lifted directly from Baldwin's map." This was certainly true for Thoreau's initial pencil tracing of Baldwin's channel. But his final, cleanly inked, version of the channel overwrites Baldwin's to portray highly detailed morphology. This detail was transferred from one-page draft maps (available in the archives) that were, in turn, drawn from field measurements taken every thousand feet along the full length of Musketaquid. Thoreau's channel resolution was also far higher than that of the opposing surveyor, John Avery, whose map was published in the 1860 report, later used for the 1861 study, and finally reprinted as a beautiful lithograph by J. H. Bufford, of Boston. This lithograph was eye candy for the court that was funded by the state legislature for the benefit of the dam owners. In comparison, Henry's map was an eyesore. Sara Luria refers to the "informal quality of the sketch and its doodle-like drawings." Though this is indeed the appearance, Thoreau's channel was in fact accurately drawn from quantitative measurements.[9]

It's at this stage of creation that Henry's scroll map gets really ugly. His measurements of channel width are the integers inked along the east bank. His channel depths were taken with a birch sapling marked off in increments of feet and inches, and written in fractions of a foot inside the channel. Aside from these numbers and their explanations, the scroll map also contains patterned areas along the full length of the river, such as stipples and hachures for "Meadows & some Hills &c by me," respectively. Additionally, there are several generations of inked comments, written in all orientations, and fainter pencil comments as well. Thoreau's final cartographic touch was to add green shading for weedy reaches, thirteen in all, many of which were located at the crossover points of successive meanders.[10]

Henry knew that the straight-line distances between bridges or survey stations used by the engineers to calculate gradient were maximum values for the actual distances along the thread of the river. As the crow flies, Musketaquid was about twenty-two miles long between the waterfalls of Wayland and North Billerica, with the main part of the valley covered by Henry's scroll map being about sixteen miles. Following the windings of the river, those sixteen miles lengthen to about twenty-five, making the river there "about ⅜ further by the windings."[11]

On July 8, Henry communicated once again with David Heard, letting him know that "I have done with your map, and, if you so direct, will leave it with Dr. Reynolds." With his base map now complete, and for the first time in his life, Henry could picture the whole of Musketaquid in a single image that was far more accurate than the prior survey maps of 1811 and 1834, the official town map by Walling from 1852, and the subsequent maps made by Avery during the 1859 and 1861 surveys. Thoreau's 1859 map is also richer in information, containing hundreds of field measurements of depth, from which he could generalize meaningful averages by reach. Finally, it illustrates and quantifies morphologic changes between 1811, 1834, and 1859, most notably the river profiles.[12]

Meadow Process

To modern wetland scientists, the historic Musketaquid meadows would be classified as either fens or marshes, owing to their rich herbaceous flora and lack of trees. Henry used the term "fen" to describe them when he was being botanical, and the word "meadow" when being colloquial. To geologists, they are floodplains, which are properly considered to be integral parts of the river. Separating the river's meadow from its channel were levees created by an unusual flood overbank process: the accretion (buildup) of torn-up meadow fragments—tufts, clumps, rafts—that were lifted by the ice during flooding, rafted down the channel, and then aligned along the bank. Thoreau called these levees a "natural brink," noting that they were generally a foot or more above the main part of the meadow. The lift-off process, Henry realized, was more common during his era than before. Otherwise, the natural meadows would not have been described for centuries as nearly smooth lawns. This was an important environmental impact being completely missed by the government committees and commissions during the hearings.[13]

Thoreau's other historical question involved meadow hydrology during the summer season. Were the meadows really wetter in his lifetime than they had been during early settlement? His instinct said yes. Specifically, he wrote that "they were probably drier before the dam was built at Billerica." From his own experience, he knew that during some summers the meadows were dry enough to be a parade ground for a "regimental muster," whereas during other summers they were persistent shallow lakes. His careful reading of Shattuck's *History of Concord* and his understanding of meadow botany strengthened his interpretation that both the meadow morphology and the botany had indeed changed, due to the flood regime.[14]

Thoreau had also noticed a recent pattern of increased woody growth on many meadows. For example: "Hardhack, meadow-sweet, alders, maples, etc., etc., appear to be creeping into the meadow," even during a wet year. To get an answer, he asked around, getting folk wisdom instead of facts: that the farmers used to mow half drunk, and therefore had the vigor to cut down swampy shrubs, whereas now that they were sober, they would cut around anything larger than a pipe stem, which might require chopping. Eventually Thoreau came to appreciate just how complex the meadow ecology was and how much it was being controlled by unseen geological factors acting differently in the Holocene versus the Anthropocene.[15]

Assabet River

On July 16, during the midpoint of his river project, Henry was mapping and sounding the lower Assabet River below the Union Turnpike Bridge as part of his geomorphology task. The flow was moderate, with a stage about half a foot above normal summer level, and rising slightly from a recent rain. Channel widths and depths were more uniform relative to the Musketaquid, averaging "about 6½ rods" and about "3¾ feet," respectively. After his previous work on the main stem, he was surprised by how broad and shallow the Assabet river was, except for one scour hole at the "hemlock eddy" that was about 8¾ feet deep." But approximately four rods "below the stone bridge," where the depth was only about 2½ feet, "there is a remarkably deep hole" for such a generally shallow stream, "up to 22½ feet below normal level." This was the deepest hole in the entire thirty-three miles of the three-river system. Having the deepest hole below the narrowest bridge was no coincidence. The bridge, being built of cut granite and only twenty-five feet wide (one-quarter of the river's average width), backed up the water as if inside a funnel and then nozzled that water eastward to scour out the hole and wash away more than a quarter of an acre of land rising about thirteen feet above the water. Thoreau then added: "All this in 35 years."[16]

This duration came from Abel Hosmer, whom Thoreau interviewed to figure out why the Assabet was behaving so weirdly. Hosmer remembered the bridge being built approximately in the year 1827, shortly after Concord's Orthodox meetinghouse was built. (Interestingly, this church was part of a sectarian war in Concord led by Henry Thoreau's maiden aunts.) Hosmer remembered the date because he had sued the Turnpike Authority for damage caused by "excessive erosion" "occasioned wholly by the bridge." His argument was that they put the wrong bridge (too narrow, "only one arch") in the wrong place (on "the regular bend") and at the wrong angle (it was "set askew or diagonally with the stream so that the abutments turned the water—& gave it a slant into the bank"). Construction may have coincided with heavy rains associated with a climatic anomaly called "false spring." From his point of view as a local landowner, the result was a disaster. For Thoreau, it was a blessing, because it made the river more dynamic and more unstable: he had long preferred this reach of the river to all others in the valley because of its sandy, pebbly, shallow, curved reaches.[17]

The excess eroded material had to go somewhere. Hence, below the deep scour pool Thoreau found a "considerable island . . . at least 3½ feet . . . above low water composed of sand." And about two to three rods below that was a truly braided river reach of "bars & islands" composed of "stones generally larger than a hens egg without sand." Below this "swiftest place on the stream thus far" was a "great eddy" where Thoreau had seen "cakes of ice go round & round in the spring." There the stream bank had been raised four or five feet above its previous level by freshets. Just downstream was a massive, actively propagating, midchannel sandbar resembling a snowdrift. He interpreted that this bar was "advancing down stream" and shedding so much sediment that it narrowed the stream farther down, and that the stream bottom was being raised by every object in its path. Thoreau calculated the volume of sediment from the eroded bank: "[One] may say that it has removed the sand to a depth of 25 feet over an area of ¼ of an acre—or say to the depth of 3 feet or 1 yard over 2 acres or 9680 cubic yards or cartloads . . . or enough to fill the present river for ¼ of a mile."[18]

The result of all this change was an Assabet that was "remarkably different" from the Sudbury. It was "not half so deep," and "considerably more rapid. The bottom is not muddy but sandy & occasionally stony—Though far shallower it is less weedy than the other. . . . This is owing perhaps not only to the greater swiftness of the current—but to the want of mud under the sand." At another point he wrote, "The very bottom of the river is loose and crumbly with sawdust." He described the "banks and bars" in this vicinity as "commonly composed of a fine sand mixed with sawdust shavings &c in which the black willow loves to grow. I know of no such banks on the main stream." Being curious, Thoreau excavated one of these banks to prove to himself that its rise, and therefore the whole of the channel, was effectively modern. Indeed, the intermixing of mineral sediment and sawdust proved that the modern river processes were due to ongoing historical disturbance.[19]

A short distance downstream were the Leaning Hemlocks, which was probably Thoreau's second-favorite place on the whole river system, after Fairhaven Pond. There, a mature forest of hemlocks on the "side hill" was being "undermined" by the Assabet's progressive undercutting. "They lean at every angle over the water," he wrote, "and almost every year one falls in and is washed away." This, he realized, was the toe of a shallow massive block that was creeping, and then caving, over the bank: "a slide and some rocks have slid down into the river." He also noted the role of the forest in holding the block together, remarking that without the interwoven roots this would have been "a flowing sand hill." Bed scour associated with these trees produced the good nearby swimming hole. With a depth of 8¾ feet, it was second only to that below the one-arch bridge.[20]

Unbelievably, Thoreau thought that the sawdust at the Leaning Hemlocks enhanced "the impression of freshness and wildness, as if it were a new country." Effectively, it was a new country created by a channel spasmodically responding to stronger flows from land clearing, enhanced erosion from bridge construction, and higher floods associated with the Billerica dam. Thoreau mapped and sounded the river in this vicinity carefully, as if he were an explorer of a new country, leaving page-sized draft maps that he never pulled together to create an extension of the scroll map.[21]

Mixed with the sawdust was a separate clue to river hydrology: "At the Hemlocks I see a narrow reddish line of hemlock leaves and, half an inch below, a white line of sawdust, eight inches above the present surface, on the upright side of a rock, both mathematically level." This sequence told a story of the river's rise, manufacturing activities upstream, the loss of hemlock leaves, and then the river's fall. Sawdust even helped reveal the eddy at the Hemlocks, "where the white bits of sawdust keep boiling up and down and whirling round as in a pot."[22]

In addition to sawdust, sand, and gravel within the channel, there was plenty of muddy and organic sediment farther out on the floodplain. Thoreau reported that Hosmer "found coal at the bottom of his meadow under the mud, three feet deep." That coal was almost certainly charcoal associated either with the initial forest clearing or with seventeenth- and eighteenth-century charcoal production, and it would likely have suggested to him that the mud above it was of historical age. Further upstream was "an Irishman digging . . . beneath black muck three feet thick" to reveal "clear white sand, whiter than common sand-hills." Perhaps Henry figured out that the mud was likely the accretion of recent sediment.[23]

Like Hosmer, Thoreau could have written his journal entry for the lower Assabet as a disaster report. He understood that the higher flows from deforested landscape, when combined with the nozzlelike effect of poor bridge design, created a cascade of effects marked by deep scour, severe bank erosion, island creation, bar migration, mud accretion, enhanced meandering, bank undercutting, landslides, the toppling of trees, and the downstream transport of a surge of sediment bulked up by sawdust. Instead, he accepted it for the lovely river it was, one made lovelier by the human impact. For the transcendentalists and their families—Emersons, Thoreaus, Hawthornes, Alcotts, Channings, and Sanborns—this disturbance gave them a shady glen with a deep swimming hole and adjacent bars to bask and picnic upon. Based on these historical associations, the Hemlocks went on to become the most popular tourist site on the Assabet. But by the time George Bradford Bartlett wrote his *Concord Guide Book 1880,* bank stabilization seems to have returned, and the living hemlocks had righted themselves back to a vertical position.[24]

THE "SAND & GRAVEL" SENT DOWNSTREAM from the lower Assabet, reasoned Thoreau, "has of course been distributed along in the river and on the adjacent meadows below." Here he's specifically writing about the Assabet's T-junction with Musketaquid, its flanking sediment bars, and the "sandy islands at the junction of the rivers." He knew that at times of strong flood flow, hydraulic ponding at the Fordway puts the bulk of Musketaquid under water, creating what is effectively a flat lake. This happened in March 1859 during the beginning of Thoreau's river project: "The town and the land it is built on seem to rise but little above the flood," he wrote. "This bright smooth and level surface seems here the prevailing element, as if the distant town were an island."[25]

Entering this slackwater from the west was the powerful flow of the Assabet River. Thoreau had watched its strong current bifurcate, with one side pouring northward down the Concord River, toward Billerica, and the other pouring southward up the Sudbury River, toward Wayland. He had watched "pail stuff" from Assabet factories float far up the Sudbury. One of his oral history informants told a story of barrels floated three miles down the Assabet from "Loring's factory" and then several miles up the Sudbury beyond Fairhaven Bay. This required that the outlet of the bay become an inlet and vice versa. From this informant he also learned that the Assabet could "rise so quickly that it forced the Sudbury to flow upstream faster than it had ever flowed down." A few old-timers told Thoreau they had seen the Sudbury flow upstream all the way to Wayland as far as "Rices." The Joint Committee saw something similar at Gulf Meadows during their May 1859 visit.[26]

Thoreau had actually felt this countercurrent when launching his boat. On windless days following strong rain, it drifted him upstream on the Sudbury. Even when the surface current was downstream, a powerful northward flow was strong enough at depth to drag large limbs upstream. Thoreau noted a line of sawdust on the banks of the Sudbury that declined in elevation to the south, indicating a former flow in that direction. He measured its southerly gradient as being "1.44 inches higher" at Nine Acre Corner Bridge than at Lee's Bridge farther south.[27]

During floods, the main Assabet current was powerful enough to scour a bedrock plunge pool at the T-junction. From there, sand and gravel in transport was carried up and out of the plunge pool to shoal both sides, creating a massive pair of flanking gravel bars. Each increment of gravel deposited by the Assabet onto these flanking bars steepens the main stem gradient to the north and shallows it to the south. Thoreau noticed this when he compared river profiles for the increment between 1827–1828 and 1859. Though the slope was a "rapid" north of the T-Junction, it was "effectively one pond above the Assabet," to the south, all the "way up to Sherman's bridge from the Stone [railroad] Bridge."[28]

To the north is the well-known Assabet Bar, marking the head of the Concord River. In this downstream direction, there's plenty of power during flood recession to regrade increments of sediment deposited on the Assabet Bar and shear that sediment downstream toward Barrett's Bar and beyond, where much of it becomes trapped. During the years of the Middlesex Canal, especially in 1835, so much sediment was pouring toward Barrett's Bar that it had to be repeatedly dredged and excavated to keep commercial scows from grounding. The fact that sawdust and wood shavings constituted a large portion of the bars in this vicinity proved to Thoreau that inadvertent human impacts from the Assabet were creating sediment pollution problems on the Concord.

To the south, at the mouth of the Sudbury River, was an even more massive, as yet unnamed bar that apparently hadn't been there in 1811. In 1859 Thoreau mapped this boat-place bar as the longest, weediest one in Musketaquid. Based on his scroll map, the end closest to the Assabet is the highest and coarsest, proving that the sediment came from that direction. He knew that, like much of Barrett's Bar, it was historical in age—at least at the top—because it wasn't noticed until after the 1827–1828 impacts. Thoreau recognized that it was acting as a dam for low-stage flows of the Sudbury River. This would have displeased his clients because it shifted the blame for the wet Sudbury meadows away from the Billerica dam.[29]

Reaches and Channels

Thoreau completed his mapping, sounding, and reconnaissance of Musketaquid and the lower Assabet on July 22. For the first time in his life, he had a holistic, systematic, and quantitative understanding of how his three rivers worked together. By that date he had worked his way downstream from Pelham Pond at the Framingham border on the Sudbury River all the way down to the Billerica dam on the Concord, giving special attention to the basin's two most important places: the natural outlet at the Fordway and the mouth of the Assabet.

His journal entry for that day is so long and so detailed that it reads like a final report for an investigation that had just ended. It reviews four weeks' worth of work, defines six distinct river reaches on Musketaquid and a seventh on the Assabet, explains how each of these reaches behaves, provides evidence for hydraulic damming at the outlet, and integrates the totality. Fifty-eight separate scientific ideas are covered in this single journal entry, which spans seventeen manuscript pages and appears to have taken two full days to write.[30]

Henry differentiated the spatial pattern of the Musketaquid channel at five different scales that zoom inward hierarchically: the overall alignment of the valley, the six specific reaches within that alignment, bedrock-controlled zigzags within those reaches, alluvial meanders and bends within those zigzags, and channel width irregularities at the scale of his boat.

At the largest scale, Musketaquid can be likened to a flat ribbon about two miles wide and twenty-two miles long between Saxonville Falls in Framingham and the Falls in Billerica. This twenty-two-mile portion of the valley spans seven linear feet on Thoreau's scroll map. The ribbon trends north-northeast along a belt of rocks recognized in the 1840s by state geologist Edward Hitchcock, whose hand-colored "Geologic Map of Massachusetts" Thoreau had examined several times. Modern geologists refer to this northeast-trending belt as the Nashoba Terrane: a slice of rock thrust beneath North America about 360 million years ago.[31]

Thoreau had long known that "Nature has divided" the Musketaquid "agreeably into reaches," which average about four miles long on the water and about a foot long on the scroll map. Recognition of these reaches, he believed, was what qualified him as a self-described "boatman." "Between Sudbury Causeway & the falls I should divide the river into these Reaches: 1st, the Sudbury Meadow or Meandering Reach, extending from just above Heard's Bridge to the Hay Bridge near the Concord line; 2d, The Fair Haven or Hill-Lake Reach, extending to Nut Meadow Brook; 3d, the Rapid Reach, which flows from near the mouth of the Assabet to Holt Ford inclusive; 4th, the Carlisle or Deep Lake Reach, which extends to the height of the hill below Jug Island, and 5th, the Billerica or shallow reach." To these five reaches from his July 22 summary he added one more from his addendum report of July 31: a southernmost reach from Pelham Pond to Heard's Bridge, which he called the Pelham Pond Reach. For each of these six reaches he provides a general description, average channel depth, type of aquatic vegetation, and so forth. Each yielded a different "sense of place."[32]

One of the things that puzzled Henry at this scale was why there were two deep lake reaches—Fairhaven to the south, and Carlisle to the north— separated by the shallowest and most rapid reach. And why was this shallowest reach—a gravelly rapids—present on both sides of the powerful Assabet? The simple answer, he later realized, is that the middle Musketaquid is an enormous deltalike mound of coarse sediment deposited by the Assabet River at its mouth, and which divides one large lake into two.[33]

Subdividing reaches were zigzags where the river was "thrown" from one place to another by geological causes unrelated to alluvial activity. These range from a mile to a quarter mile in length on the water and a few inches on his map. "It is remarkable," Thoreau wrote, "how the river (even from its very source to its mouth) runs with great bends or zigzags regularly recurring & including many smaller ones." He recognized that these were places where the thread of the river was forced to flow either parallel to the tectonic grain or across it within a younger geological fault zone. Henry compared these large bends to the "wriggling of a snake controlled between 2 fences." Overall, he saw this pattern weaken toward the north, where the river became more direct and simple, as "if a mighty current had once filled the valley." This was a prescient insight. Indeed, during final stages of the last glaciation, powerful meltwater streams drained southward beneath the ice sheet and then northward toward it.[34]

For Thoreau, the most spectacular of all these zigzags lay between Fairhaven Pond and Ball's Hill at the north end of Concord's Great Meadow. Between these end points the river zigs northwesterly, leaving the ancestral valley to flow through the river's most rocky reach between the cliffs of Conantum and Fairhaven Hill. This subreach follows the trace of a geological fault zone that cuts the prevailing layering. Upon reaching Nut Meadow Brook, this subreach zags sharply to the northeast to follow the bedrock strike all the way down to Ball's Hill, where it rejoins the ancestral valley. This zigzag forms two sides of a triangle whose hypotenuse is the chain of kettle lakes aligned with and including Walden Pond. There the original valley became plugged with sediment during glacial retreat, burying residual ice blocks that later melted to form the Concord Lake District.

Smaller than the zigzags were alluvial meanders best expressed within Sudbury Meadow. These natural river bends—also present in the lower Assabet—were independent of the bedrock or the glacial geology, and were instead caused by helical or corkscrew-like flow in a naturally curved channel. On the ground they were less than a quarter mile long, and on the map less than an inch. Meanders, he noted, developed where the gradient was low and the bank cohesive.

Even smaller than meanders were channel irregularities at the scale of his boat, perhaps a large erratic from the diluvium, a scour pool, a patch of weeds, or the hollow produced by a bankside tree. On his map, these features span only an eighth of an inch or less. Yet they were carefully drawn and often annotated.

THOREAU HAD KNOWN ABOUT THE REMARKABLY low slope of the Musketaquid channel for many years. During his 1859 river project, he calculated it to be "just 1½ inches to a mile," with most of that drop taking place in Concord. He couldn't resist a private joke: "If Concord people are slow in consequence of this river's influence—the people of Sudbury & Carlisle should be slower still." The channel depths he measured were trivial compared to channel lengths, yielding a ratio of "about 1/17,000." For an analogy, he calculated this ratio as equivalent to the thickness of two ordinary sheets of paper laid on his seven-foot-long map. The thickness of four sheets of paper, he calculated, would take in all the "deep holes which are so unfathomed & mysterious not to say bottomless to the swimmers & fishermen." These comparisons parallel those he made of the depth-to-width ratio of Walden but are two orders of magnitude smaller.[35]

The size and shape of Musketaquid's channel varied as much as its pattern. Using an eleven-foot depth as an arbitrary cutoff, he identified twenty-seven deep holes in Musketaquid, nine of which were inherited from geology and nine of which were created by alluvial processes. In general, depths increased to the north. Using a four-foot depth as an arbitrary cutoff for shallows, he identified twenty-three "shallow places" over the same length.

Twenty-one of these were "weedy places," defined as places where weeds "extend quite across" the entire channel. Their presence signaled a sluggish current. When "clogged with weeds. . . . the water weeps, or is strained through" the channel like a sieve.[36]

In his July 22 report Thoreau made sure to emphasize Musketaquid's superlatives. The shallowest place was at the Fordway, where the water was "generally only 2 feet—with a hard bottom & numerous rocks in its bed. It is quite fordable in a carriage." The deepest, at 19½ feet, was "Purple Utricularia bay," near the entrance to Fairhaven Pond. Narrowest was at Farrar's Bridge, near the "SW line of Concord," where the river measured only 53 feet wide (he used the convention that the channel begins when the water is one foot deep). Excluding Fairhaven Pond, the widest pace was on the Carlisle reach between "Squaw Harbor & Skelton bend." The weediest place was at the Sudbury Causeway because that shallow bend had been completely cut off by the Sudbury Canal. The sandiest was Barrett's Bar. The muddiest were "the most stagnant parts" up and down the river. In his journal Thoreau linked all of these appearances—shallows, deeps, narrows, widths, weeds, sand, mud— to their respective geological and hydraulic causes, namely, bends, currents, shoals, banks, constrictions, expansions, and variations in sediment texture.[37]

Of the many shallow places, four were those where the river encountered bedrock obstructions: Hubbards Bath, Egg Rock, Middlesex Turnpike Bridge, and the Falls. All coincide with changes on the state's current official bedrock map, confirming Thoreau's powers of observation. The most important of these bedrock shallows was the Fordway and the rapids immediately below it, where the combined flow of the main stem drains through a notch cut slightly down into the bedrock lip of the Musketaquid basin by subglacial streams. There Thoreau found many "lumpish boulders" carved by flowing sand and pebbles into the shape of low pedestals that were stained "black as ink" below the water. He interpreted this as having once been the bed of a rushing gravel stream like those he had seen in the Maine wilderness. This is indeed what had happened during the transition between the Pleistocene and Holocene epochs.[38]

At the Fordway, the solid granite bedrock in a long channel of low slope was very resistant to being widened or cut down, and the roughness of its "lumpish" boulders forced the water to flow above and around these obstacles like a crowd of people backing up behind turnstiles in a subway station. During each flood, this basin outlet became a hydraulic dam for an enlarging transient lake upstream. And because the Musketaquid basin held such a great volume of water, these temporary lakes lasted for weeks, plenty long enough to fertilize the meadows and recharge the flanking aquifers.[39]

The earliest Puritan colonists recognized the Fordway as a serious impediment for free drainage, which is why they tried to excavate the channel there as early as 1636. This historical fact was the ace in the hole for the respondents of the 1859 flowage controversy, because it proved that persistent problems with upstream flooding predated construction of the first gristmill by a century, and the stone dam by two centuries. At the Fordway Thoreau noticed the rubble from early attempts to deepen it. He also noticed the underwater archaeology of those who used it as a crossing, noting the "bricks & white crockery" associated with countless bumpy river crossings dating back to the earliest colonial times.[40]

The remainder of Thoreau's river shallows were all sediment shoals. And all but one were located where they should be, downstream from bends or in places where flow was expanding rather than contracting. The sole exception was Barrett's Bar, in Concord. To Thoreau, it was the most "conspicuous" bar in the entire river for "shallowness & length, all-together." In modern terms, it was an anomaly of location, length, channel form, and sediment composition. It's the only distinctly multibranched part of the river, with two perennial channels flowing around a high island of sediment between them. Bedrock likely has some influence at depth. Thoreau described this as "a narrow channel on each side deepest on the south," speculating that "if a large piece of meadow should lodge on this it would help make an Island of it rapidly." River geologists have since discovered that such places are diagnostic of streams that are overloaded with sediment relative to stream power. Indeed, Thoreau noted in 1859 that Barrett's Bar is "composed in good part of Saw dust—mixed with sand," and he concluded that it was a product of the Assabet River sediment surge. During his era, it was the first significant rise in the river surface profiles above the Billerica dam, one that remained in place even during floods. Below it, the river was effectively a long lake.[41]

10 | Genius

THOREAU'S PROGRESS REPORT OF JUNE 22, 1859, CULMINATES with a stroke of genius that explains why Musketaquid is so prone to flooding and why its meadows were so fertile. He realized that the alluvial bed of the river for at least twenty miles south of the Fordway lay at or below its bedrock outlet. This showed Musketaquid to be a single bedrock basin between Billerica and Saxonville that held a single lake during times of high water. The meadows occupied wide spots in the basin, and the narrows were places where resistant Pleistocene or Paleozoic materials crowded the valley from the sides. Each meadow was like a flat bead on a necklace lying flat on a horizontal table. Each narrow was like an exposed piece of the string.[1]

Thoreau's induction generated a half dozen thought experiments that he tested with his depth soundings. At Walden Pond, a similar data set led to his insight of radial symmetry, what he called the "law of two diameters." For Musketaquid, he realized that "a river of this character can hardly be said to fall at all—it rather runs over the extremity of its trough—being filled to over flowing. Its only fall, at present . . . is like the fall produced by a dam—the dam being in this case the bottom" of the Fordway. After twenty-five miles of flow, "it cannot be said to have gained anything or have fallen at all. It has not got down to a lower level. You do not produce a fall in the channel or bottom of a trough by cutting a notch in its edge. The bottom may lose as much as the surface gains."[2]

In a northward (downvalley) direction, "I should say that there was more of a rise than a fall" between the bottom of the river in Sudbury and the Fordway. Taking this insight to the next level, he recognized the valley as so lakelike that "if our river had been dry a thousand years it would be [difficult] to guess even where its channel had been." These insights led to additional inferences and observations: "The 3 deepest places measuring from the surface of the water—are also the deepest absolutely," and "the deepest point is far to the south of the outlet, opposite what it should be." His use of the word

FIGURE 21. Leaning Hemlocks. Historic photograph by Wendell Gleason. Taken from opposite bank, Concord, Massachusetts, 1899. Note evidence for active bank erosion and still water. Courtesy of the Concord Free Public Library.

"opposite" here brings his thinking very close to what we now know: that the crust has been tipped downward to the south, opposite the direction of river flow.[3]

Thoreau's observation that the deepest holes lay farthest upstream led him to deduce the actual mechanism by which the Concord River keeps its channel open. It's a style of circulation similar to that of the Mississippi River. In Musketaquid, a very slight south-to-north gradient at the top drives water northward at the bottom due to a pressure gradient. At the bottom, he wrote, the water can flow uphill "given the overall direction of the valley." And sediment can be moved northward up the slope provided the small loss of potential energy over a vast width is concentrated in constrictions to create a current strong enough for basal sediment transport. This is how the Mississippi River moves its bottom sediment for its final 450 miles. There the bed of the river lies below the level of the sea into which it drains. The slight surface slope of about 3 inches per mile creates an internal "tumbling effect as water spills over itself, like an enormous ever-breaking wave."[4]

The whole week had been extremely exciting for Thoreau. His hydraulic insights from July 23 spilled into July 24, reappeared on July 30, and then gushed so strongly on August 3 that his journal reads like an addendum to an addendum to an addendum. "The river is but a long chain of flooded meadows," "a mere string of lakes which have not made up their minds to be rivers—as near as possible to a standstill." "Though the current can be felt in shallow and narrow places, it faints & gives up the ghost in deeper places—. . . & you would presume it dead a thousand times, if you did not apply the nicest tests—such as a feather to the nostrils of a drowned man." The "swifter places," he wrote, "are produced by a contraction of the stream—chiefly by the elevation of the bottom at that point—also by the narrowing of the stream." The Fordway, located on the lip of the basin, was swiftest because the drainage from more than 300 square miles was being concentrated into what had likely been a subglacial notch.[5]

Having dedicated himself to finding and reverencing the causes beneath the appearances of any "created thing," Thoreau reached back into geological time to explain the whole of Musketaquid. The three deep lakelike reaches he had identified—Pelham Pond, Hill-Lake, and Carlisle—"may, perhaps be considered as 3 deep holes made by a larger river or ocean current in former ages." Indeed, these were places of extra glacial scour, located where zones of rock weakness along brittle fault zones were roughly parallel to ice flow rather than transverse to it.[6]

AT THIS STAGE, THOREAU HAD MAPPED and fathomed and measured and explained all three of his rivers. Reading this section of his journal a reader might conclude he was reaching the end of his river project and was preparing to return to his default summer practice of botanical inventory. But shortly after writing his July 22 report, he encountered "Simonds of Bedford . . . measuring the rapidity of the current of Carlisle Bridge" using a board on a string. That encounter seems to have inspired Thoreau to launch his own research program in channel hydraulics.[7]

Thousands of journal entries from the previous ten years show that Thoreau was familiar with every component of watershed hydrology: the gravitational pull of liquid water through the watershed, the pull of precipitation down from the sky, the pull of that water down into either rills or aquifers, the pull of those aquifers into springs and brooks, the pull of those brooks into tributary rivers, and the pull of those rivers through the sluggish main stem. Never before, however, had Thoreau performed a systematic program of measurements of river currents and velocities. Inspired by Simonds, the very next day he began to make a series of twenty current calculations in nine places, most of which were in the lowermost Sudbury near his boat place. This work, when combined with his discharge estimates for the compensating reservoirs, was an achievement unmatched by anyone else during the flowage controversy.

To estimate velocity, he launched a variety of objects into the water and timed how long it took for them to move a specified distance, usually 100 feet. For surface measurements he used wood chips and a board. For subsurface measurements he used a bottle filled to neutral buoyancy. To integrate local turbulence over a larger cross-sectional area, he submerged a weighted strawberry box. He also made qualitative observations about the strength of flow at the bottom, based on the angle taken by rooted vegetation in the direction of current. Using various combinations of measurements and observations, he showed how current velocity changed from river bottom to top, and from riverbank to channel center. His measured velocity profiles match both modern theory and practice.[8]

Working like an oceanographer, he also differentiated a wind-driven surface current from steadier pressure-driven flows. Using his integrated velocity measurements, he then computed the travel time for hypothetical slugs of water moving down the main channel from Wayland to Billerica. This process, called flood routing, is the basis for flood forecasting today. His estimate of forty-eight hours' travel time for the sixteen miles of straight-line distance for the alluvial valley lies well within the range of error for modern studies, especially when you take into account the increased length along the curved channel.[9]

Modern channel hydraulics confirms his summary of observations: "I should infer from this that the swiftest & most uninterrupted current under all conditions was neither at the surface nor the bottom, but nearer the surface than the bottom—If the wind is down stream it is at the surface—if up stream it is beneath it, and at a depth proportionate to the strength of the wind. I think that there never ceases to be a downward current." In a further analytical step, he combined different channel cross sections with different current measurements to prove that current velocity everywhere was directly proportional to the cross-sectional area: "The sluggishness of the current, I should say, must be at different places, as the areas of cross sections at those places."[10]

By the end of his first day of hydraulic work, Thoreau had calculated a river discharge, which is the volume of water moving through the channel per unit time at a particular place. Using his best trial for stream velocity at the boat place (an average of "100 feet in 4½ minutes"), a width of 88½ feet based on his 1-foot-to-1-foot rule, and the pattern of depth soundings to give him a channel cross section, he calculated that roughly "266 cubic feet pass here in a minute." Such discharge estimates are challenging calculations because they involve channel area, velocity, and roughness. Much simpler is river stage, which is little more than a dipstick measurement of water height. Establishing the mathematical relationship between these two parameters—stage (simple) and discharge (complex)—is a primary objective of modern hydrological studies because the easy measurement can be used to predict the challenging one. The result is called a rating curve. Though Henry never produced one, he theorized every step of the process.[11]

THE COMPENSATING RESERVOIRS NEAR THE WATERSHED DIVIDES of the Sudbury River in Hopkinton and the Assabet River in Marlborough were enormous, flooding nearly a thousand total acres. In Henry's river project file at the Concord Free Public Library is an undated report titled "The Overflow of the Meadows of the Sudbury River—Report from the Board of Cochituate Water Commissioners." On it he penciled "Aug 59," proving that his private river project coincided with the public flowage controversy long after his paid consultancy for the River Meadow Association in June had ended. This report announced that after a "full review of the facts the Water commissioners come to the following conclusion": that the summertime floods and meadow wetness preventing crop harvests were due to "an imperfect natural drainage, and this natural drainage is by natural causes continually becoming more imperfect, without any regard to the artificial ones superadded." In short, and without any evidence, they wrongly dismissed the claims of the petitioners that the compensating reservoirs were causing summer flooding.[12]

Thoreau actually tested the Commissioner's unfounded assertion by calculating—in pencil on a loose sheet of notebook paper—the discharges from the compensating reservoirs. He found that the combined discharge released from the reservoirs was significant, averaging 17,823,180 gallons per day, or 2.6 times more than the natural outflow from Lake Cochituate. For the Sudbury River, these reservoir releases were 2.4 times stronger than his measured river discharge at the boat place bar. The rough equivalence between these independent calculations confirms his methodology and proves quantitatively that meadow wetness would have been increased behind weedy bars, which retarded the flow. No one else made such an important calculation, information that would have been extremely helpful to his former clients.[13]

To his quantitative measurements of current velocity and discharge, Thoreau added qualitative estimates of flow velocity at the outlet, where it was fastest, and at Nine-Acre Corner, where it was most sluggish. The latter was truly a lake, with no detectable current at all, not even at the bottom. The former was like a canal on June 22: he found that he could easily paddle against the current in the Fordway's shallow, rough bedrock channel. For that to be possible in a stream draining hundreds of square miles required an exceptionally flat surface gradient, which in turn required that the Fordway be backed up by the dam below, an observation he would later confirm. On June 24 he learned that the current in the nearby Carlisle Reach, based on velocity measurements, was "swiftest" just after the water had begun to fall. This quantitative observation proved that hydraulic damming had occurred at the Fordway because the transient lake was being unplugged from below.[14]

Hydraulic Experiments

On July 27 Henry had a friendly chat with John Avery, the professional surveyor for the respondents. At the time, Avery was midway through the gradual process of surveying the full length of Musketaquid to determine how its surface elevation changed with distance from the dam. Beginning with a backshot to the iron bolt at the dam, he swiveled his "leveling instrument" around to shoot forward up the river as far as he could see. Then, shooting back to that same upstream point, and repeating the process, he carried that survey line all the way up to Bridle Point Bridge in Wayland. This allowed him to know the height of the water surface at any point relative to the legally mandated level of water in the dam pool. Commencing July 13, Avery's leveling survey took him about three weeks to cover the twenty-two valley miles.[15]

Thoreau encountered Avery in the vicinity of Concord's South Bridge. There Avery told him that he was finding a greater change in elevation than the 1834 Baldwin map had indicated, which was bad news for Thoreau's colleagues, the petitioners. This free exchange of information likely would not have taken place had Thoreau revealed to Avery that he had been a consultant for the petitioners, who opposed the respondents employing Avery. This surveyor to surveyor chat is circumstantial evidence that Henry was being covert, perhaps even "spying" on the opposition. Thoreau had had at least one previous encounter with Avery. On that earlier trip Avery had been marking the instantaneous water level on bridge abutments, in preparation for measuring the elevation later. To accomplish this task, he traveled quickly upriver over bumpy roads in a horse-drawn chaise. That Thoreau encountered Avery on both trips indicates he was paying very close attention to the river science.[16]

The respondents were financing Avery's work in the hope of resolving the discrepancy between the 1811 and 1833 surveys in their favor. This new survey would also provide a horizontal baseline for a series of scheduled drawdown experiments to be conducted at the Billerica dam. Their experimental model was to quickly drop or raise the level of the dam pool and then monitor how the upper surface of the river at distant points—its water-surface profile—responded to this known change. Theoretically, an instantaneous drop at the dam pool should propagate upstream as a transient wave of enhanced flow that will eventually stabilize under the new conditions. Conversely, a rapid rise of the dam pool should propagate upstream as a wave of reduced flow. To resolve actual spatial patterns of stage changes in the Musketaquid, all local changes had to be referenced to the same common datum, in this case the iron bolt at the dam. Though this program began with a preliminary experiment on July 12 by dam superintendent Israel Colson, all the good data used for the court proceedings depended on Avery's new survey.

Knowing Avery's plan, Thoreau set up his own parallel program of stage monitoring, almost certainly to privately verify or refute the official experiment. On July 25 Thoreau reset his local datum by pounding a stake into the riverbank, locating its horizontal coordinates on his scroll map, and letting its vertical height be "X." For convenience, he notched a willow at his boat place at the same height.[17]

On July 31, under normal summer conditions (base-flow), the gates of the Billerica dam were opened to allow the pool height to fall three feet. On that same day, Henry began his program of four-times-daily stage measurements (typically near dawn, midmorning, midafternoon, and dusk) and noted the stage to the nearest $1/64$ inch, a level of precision greater than was used in the official investigation; he would extend these measurements through the morning of August 15. Between August 4 and 8, the chief engineering consultant for the factory owners, James Francis, conducted the first of his official monitoring experiments.

For the duration of this experiment, Thoreau's journal contains little but river stage data, indicating his intense interest in the process. The August 5 entry overlapped with the initial results of Francis's experiment and is rich in text, and therefore serves as a preliminary report. Though the dam pool had been drawn down low enough to dry the canal completely, Henry noticed from Avery's benchmark that the water level at Concord's North Bridge (Flint's) didn't drop more than an inch and a half. In the next phase of the experiment, Talbot let the water rise back up some four feet, again with little change in Concord. Thoreau's overall conclusion was that, under low-flow conditions, great changes in the height of the Billerica dam had minimal effect on the Concord River above Barrett's Bar.[18]

Thoreau's next step, on August 11, was to extend this local datum on the lower Sudbury River to other sites, notably at the Leaning Hemlocks, Hunts Bridge, Woods Bridge, and the Eddy Bar. Of these, only the station at the Hemlocks gave him a data set useful for comparative purposes. Though of limited duration, this data set provided compelling evidence that Henry's main datum on the boat-place bar was more conservative than its counterpart on the lower Assabet, which rose earlier and higher, and fell faster and lower.[19]

Henry summarized both of these drawdown experiments on August 14: "In the course of the above 16 days, very great changes are said to have been made at the Billerica dam—but I have seen no effect of any here." This conclusion, coming only nine days before local testimony would begin at the Concord courthouse, flatly refuted the expectations and hopes of the petitioners, who were his clients during the early consultancy phase of his river project. Thoreau's data remained safe in his private journal.[20]

AN UNEXPECTED OFFSHOOT OF HENRY'S OBSESSIVELY detailed monitoring of the drawdown experiments was proof that the human presence in watershed behavior was pervasive. With the river at a "low water . . . standstill" and being "unaffected by rain," he noticed the river stage toggling above and below his stake by about $13/16$ of an inch each day. Because it fell at night and rose in the forenoon, he called it "a regular tide." Realizing that the water rose higher on Sundays and fell much more quickly on Mondays, he knew that this was not a natural phenomenon but a consequence of human action. "By a gauge set in the river I can tell about what time the millers on the stream & its tributaries go to work in the morning & leave off at night & also can distinguish the Sundays—since it is the days on which the river does not rise but falls. If I had lost the day of the week I could recover it by a careful examination of the river. It lies by in the various mill-ponds on Sunday & keeps the Sabbath. What its persuasion is, is another question."[21]

To supplement his stage measurements at discrete stations, Henry also carefully studied the patterns of sawdust accreting on the riverbanks. This was similar to what he would later do with tree rings, emphasizing salient lines, overall trends, and distinct gaps. The sawdust, being fed into the stream from the Assabet Mills, acted like a tracer. And because it floated and adhered to the bank, it created perfectly level lines that were far more accurate than a "coarse chalk-line made by snapping a string." The sharpest lines indicated stage reversals from rising to falling, similar to the lines of flotsam at the height of the spring freshet and of the seacoast tide. Dull lines indicated decelerations in falling stage. Multiple lines formed a pattern of "very distinct parallel lines four or five or more inches apart," similar to those of a bar code. These allowed him to visually compare stage behavior for all three rivers near the triple point. What he learned was that river stage was far more interesting and sensitive than was being revealed, even by his own highly quantitative data sets.[22]

Each stream had its own schedule of rise and fall, depending on water sources and on whether its mouth was affected by the flood of a larger stream. The Assabet, for example, would crest within a day of a heavy rain, whereas the main stem always took at least two days. Data from Henry's monitoring from 1860 would show that it took three days for the lower Concord River to crest after a heavy one-day rain in June. This three-day lag time was consistent with what he had learned from an old-timer named Flood: "Old settlers say this stream is highest the third day after a rain." This was also consistent with his calculations of travel time for a slug of water to work its way through the full length of Musketaquid from Wayland to Billerica. Here were three lines of evidence supporting one another: empirical data, qualitative history, and predictive calculation, respectively.[23]

Thoreau understood the details as well: that the timing of flood waves "depends on the amount of the rain, the direction and force of the wind," and the time of year. Indeed, "a southeast wind will take the water out sooner, and any strong wind will evaporate it fast." And because the valley parallels the prevailing southwesterly winds of summer, conditions then are "very favorable to its rapid drainage at that season." During June "the water is usually falling," meaning that "even a heavy rain might not raise the river as much as it falls each day." All in all, his private, self-funded scientific methodology was more robust than that of anyone else in the public, government-funded study.[24]

Vitesse de Régime

The most intellectually thrilling part of Thoreau's entire river project involves his trip to the Harvard Library in Cambridge on August 15. This was the first day following the conclusion of Francis's drawdown experiments, giving Thoreau the liberty to leave town. Likely he was browsing the stacks, looking for a resource to help him understand the movement of sediment bars on the lower Assabet and on the uppermost Concord below it. On that day, he checked out *Principes d'hydraulique et de pyrodynamique: vérifiés par un grand nombre d'expériences faites par ordre du gouvernement,* the most influential work of channel hydraulics from his era. Its author, Comte Pierre Louis Georges Du Buat, was a lieutenant colonel in France's Royal Corps of Engineers. Being from a wealthy Normandy family with royal titles, Du Buat had been forced to flee with his family in 1793, during the French Revolution. His work was central to the French School of channel hydrology, which was then the most advanced engineering program in the world and served as the model for the U.S. Army Corps of Engineers. Du Buat was especially interested in the open channel flow of natural rivers.[25]

Originally published in 1779, Du Buat's work was republished in an expanded three-volume third edition in 1816. Reading its first volume in French, Henry translated seventeen pages of longhand notes that were "laid inside Thoreau's manuscript journal for April 8–September 21, 1859," as a "letterfold," meaning it was considered part of his journal. Because Thoreau wrote the date "Aug 9" in pencil on the top of those notes at an angle, both Kenneth Cameron (a scholar of Thoreau's notebooks and correspondence) and the Princeton transcription team assigned that date to the notes. But Thoreau didn't check the Du Buat work out of the Harvard library until the fifteenth of August. Because library records are more fastidious than Thoreau's memory, and because the date of August 15 dovetails with Thoreau's actual activities, I conclude that his work on Du Buat's text closely postdates the fifteenth, and precedes the twenty-third, when Thoreau returned to his natural history pursuits. He may have written "Aug 9" on the letterfold because it was a critical day of the drawdown experiment, or because that day's journal involved an explanation of channel curvature.[26]

I've found no indication that Du Buat's work on sediment transport, river equilibrium, and channel process was being used by any other natural scientist of antebellum America. In that era, such quantitative experimental and theoretical work on river channels was considered the intellectual territory of civil engineers. Du Buat, however, was clearly doing the same sort of Enlightenment natural science as his equally aristocratic French contemporary Antoine Lavoisier, who named the element oxygen in 1775. Both were seeking quantitative laws for natural phenomena—in Du Buat's case, laws that would predict the shape and character of natural river channels. He was especially interested in what he called the *vitesse de régime,* or steady-state equilibrium between form and process.[27]

Thoreau's seventeen pages of handwritten notes prove that he was hopelessly and happily trying to understand the river as a "lawful creature," to use Emerson's phrase. More specifically, to explain how each specific reach or channel segment worked as a system of checks and balances through negative feedback. Thoreau's translation emphasizes the powerful idea that some ideal channel shape "secures the greatest swiftness when the inclination is constant, or which requires the least inclination if the swiftness is given." This was heady theoretical stuff. More specifically, Henry's journal reveals that he had independently integrated the four most important variables in channel morphology: the velocity of flow, the slope of the channel, and the frictional resistance offered by two variables, channel shape and roughness. Thoreau's journal sketch for an archetype river channel exactly matches Du Buat's text: "the figure of a trapezium in which the breadth at the bottom is ⅔ the depth of the water & the slopes are ⅘ of this depth." Finally, he copied notes for a subject that had intrigued him since childhood: river curvature. Du Buat's work confirmed his earlier inductions that meander migration was both a cause and a result of helical flow.[28]

As a related part of this work, Thoreau hand-copied a table of data linking sedimentary particles to the threshold velocities needed to transport them, based on controlled experiments. This ranged from 0.25 feet per second for soft clay to 4.0 feet per second for "heavy shingle" gravel. He also translated Du Buat's correct theory of ripple formation by a process called saltation.[29]

During his library research into sediment transport, Henry also checked out *Appleton's Dictionary of Machines, Mechanics, Engine-Work and Engineering* from the Concord Free Public Library. Robert Sattelmeyer concluded: "T's extracts are from the article 'Hydrodynamics' and have to do with the velocity and carrying power of river currents." This reference suggests Thoreau was seeking a second opinion to back up his conclusions from Du Buat's research.[30]

DRIVING THOREAU FORWARD THROUGH THIS VERY challenging reading was his need to know why alluvial channels look the way they do.

The answer known to modern fluvial geomorphologists involves a simultaneous force balance and mass balance at a threshold discharge called bankfull flow. Geologically, not much happens to rivers during low-flow conditions, when they effectively become groundwater drains for H_2O and dissolved ions. But during bankfull conditions, the size, shape, slope, bank, and bed of channels become interdependent variables that are constantly adjusting to one another. This give-and-take occurs mainly through the addition and subtraction of sediment by erosion in one place and deposition in another, and by the adjustment of surface slope to sediment texture. One result is the migration of sand ripples, sediment bars, and channel banks. Another result is predictable change from boulders to sand. Still another is the regular pattern of scour and fill on the riverbed. During these channel-forming flows, each reach of a stream is like a "lawful" creature adapting to external changes via internal adjustments in an attempt to stay as close as possible to the *vitesse de régime*.[31]

Ultimately, all river sediment comes from the land. It reaches streams externally through rills, ravines, and landslides. Once delivered to the stream, it is stored as floodplain sediment that will later be mobilized by bed and bank erosion. For Thoreau, Clamshell Bank provided a wonderful example of the linkages. In February 1855, he personally witnessed the creation of a "ravine some ten feet wide and much longer," which he predicted would "go on increasing from year to year without limit" through positive feedback as a consequence of the human makeover. Forest denudation on the hillside had enhanced both seepage and surface erosion. Persistent high water had weakened the soils by raising the pore water pressure. The larger the gully became, the more seepage it gathered, and the faster it grew, and so forth. Thoreau did not interpret this "ravishment" as blight. In contrast, it became an archaeological treasure trove, a special habitat for hundreds of mosquito-eating bank swallows, and a source of sand for improving the clam habitat. On March 19, 1859, he recognized a new ravine there that gave a "new & remarkable character" to the landscape, "as does the deep cut on the RR." Its "strange & picturesque" appearance helped culminate Thoreau's river project, as did the sand foliage at the Deep Cut did for *Walden*.[32]

Such observations led Thoreau to his general theory of valley creation and to an explanation for how large rocks came to be found in the midst of meadows: "Rivers are continually changing their channels,—eating into one bank and adding their sediment to the other,—so that frequently where there is a great bend you see a high and steep bank or hill on one side, which the river washes, and a broad meadow on the other. As the river eats into the hill, especially in freshets, it undermines the rocks, large and small, and they slide down, alone or with the sand and soil, to the water's edge. The river continues to eat into the hill, carrying away all the lighter parts [of] the sand

and soil, to add to its meadows or islands somewhere, but leaves the rocks where they rested, and thus in the course of time, they occupy the middle of the stream and, later still, the middle of the meadow." This is how Saddle Rock in Wayland, the longest-duration hydrological datum in the valley, came to be located in the middle of an alluvial valley, where it was later put to good use.[33]

During his study of the heavily impacted lower Assabet, Henry noted that channel migration within the floodplain had consumed as much as 50 rods (825 feet) of the former riverbank, much of it in response to human disturbance. The corollary to this observation, he realized, was that much of that missing mass had been incrementally deposited on the opposite bank of the next bend downstream; otherwise, streams would be infinitely wide. "It is remarkable how the river—while it may be encroaching on the bank on one side, preserves its ordinary breadth by filling up—the other side. . . . There are countless places where the one shore is thus advancing [and], as it were, dragging the other after it." Meander migration of such high magnitude and rapid rate is usually an indicator of human disturbance.[34]

Once entrained or incorporated, sediment migrates downstream according to its particle size. In Henry's emerging theory are these statements: "The heaviest particles of alluvium are deposited nearest the channel," "The architect of the river builds with sand chiefly not with mud," and "Mud is only deposited very slowly in the stagnant places—but sand is the enduring building material." Though the bottom of the Concord River is muddy in many places, its "bottom is occasionally. . . . of soft shifting sand ripple-marked . . . under 4 or 5 feet of water." For clean sand, his last observation can be used to estimate a basal drag velocity of approximately two to three feet per second. Such ripples, especially those in Nut Meadow Brook, fascinated Henry throughout his sojourning life. Like the flowage of the fine sand and clay at the Deep Cut, ripples were an emergent phenomenon composed of inorganic material, but sharing some properties with life. Like the clay at the Deep Cut, ripples straddled the continuum between the mineral and vegetable kingdoms.[35]

The narrative of Thoreau's initial study of the Assabet (mid-July), his reading of Du Buat (mid-August), and his theory of channel form and migration (late August) illustrate his method during this final stage of his scientific career. Prompted by field observations, he took advantage of easy access to public transportation, his alumni borrowing privileges at a research library, his facility with multiple languages, and his habit of transcribing original sources to self-educate in a subject he previously knew little about. The result was a sophisticated understanding of the Concord River that surpassed that of everyone else involved in the flowage controversy. In that moment Thoreau may possibly have known more about rivers as "lawful creatures" than anyone else in America.

Daily Traveller

Two days after Henry's August 15 visit to the city, the *Boston Daily Traveller* published an article titled "Report of the Cochituate Water Board on the Petition of Proprietors of Meadows on the Sudbury River," which was identified as "City Document No. 49." It was authored by someone who signed his name only as "R." This report strongly opposed the Billerica dam, described the geological setting and origin of the Concord Valley, and recommended not only tearing down the dam but also blasting and excavating the Fordway so that the valley might become an agricultural paradise.

Henry may have been an anonymous source for the article. One argument involves timing. The article came out two days after Thoreau's first trip to Boston of the summer. This would have given "R." one day to write it, and another to have it published. Second, the vitriol spilled on the factory dam owners is consistent with Thoreau's bias. But the strongest argument for Thoreau being a source involves the article's description of the Musketaquid valley. Its text matches Thoreau's unpublished subdivision of Musketaquid's reaches, gives their proper lengths, and uses geological concepts present in Thoreau's private July 22 report. This speculation is supported by the singularity that this newspaper article is the only one in Thoreau's file, even though many others must have been published that summer.[36]

AUGUST 17 MARKED THE ONSET OF THE SECOND official drawdown experiment at the dam. On that same date Henry returned to his independent river stage monitoring, which he carried on with through August 22. That day his data caught the nadir of the 1859 drought. Fortuitously, this was only one day before the Joint Committee returned to Concord to view the river at its low-water stage when the meadows should have been driest, and to begin local hearings. The timing of Thoreau's activities this summer, especially with respect to the drawdown experiments, shows that he had effectively become a silent third party during the flowage controversy, a solitary scientific genius enjoying the most challenging research project of his life.[37]

11 | Saving the Meadows

AFTER MORE THAN A MONTH OF LEGAL PROCEEDINGS IN BOSTON, the Joint Committee returned to Concord on August 23, 1859, and toured the river the following day. Their visit was scheduled to coincide with late summer's low-water conditions. Based on Thoreau's stage monitoring, their timing was perfect. In contrast to the high water of mid-May, the river lay well below its banks throughout the valley at the lowest stage of summer.

This time the entourage went north and south on the same day, "sailing in a steam-tug the whole length of the stream from Farm Bridge to Billerica." This required a round trip south to Wayland and back, followed by a round trip north to Billerica and back, the same sixty miles that Thoreau had skated round trip in a single day many years earlier. In Sudbury they reported the current to have been exceedingly sluggish, its flow having been nearly stopped by weeds growing in patches up to half a mile in length over various sandbars such as Bent's Bar in Wayland. Churning through the weeds must have been challenging. In the northerly segment, Barrett's Bar below the Assabet was seen as the greatest of all "natural obstructions except that of the Fordway," which they saw as "an obstruction to the passage of water almost as great, apparently, as the dam itself."[1]

Upon arriving at the Billerica dam, the petitioners suggested "that the water should be drawn down, in order that the Committee might see the old Dam and ascertain its relative height." They were referring to the original 1798 dam for the Middlesex Canal, which was submerged by the higher stone-faced dam of 1828 built immediately downstream. Both parties engaged to do so "upon some Saturday night which parties might agree on," provided that the Joint Committee was notified. Saturday night was best for a drawdown because the river would have the Sabbath to recover in time to turn factory wheels on Monday morning.[2]

The resumption of hearings on August 23 coincided with Henry's return to his normal sojourning practice. On that day he provided his first detailed record of natural history observations since July 12, when the passage of the surveyor John Avery through town inspired his stage monitoring work. For the next three days, Thoreau's journal returned to his old groove of musing about natural history and philosophy. Beginning August 26 its entries read like those of early summer prior to his intense engagement with the river project. Things went smoothly for the next two weeks, until September 16, which Thoreau spent surveying along Bedford Road for Ralph Waldo Emerson. Then, for the next month, Thoreau's journal entries brimmed with rants against superfluous wealth, the immorality of slavery, annoying music, social niceties, capital markets, and abolition politics. The river project, at least for now, seemed a thing of the past and his journal no longer a scientific notebook. During this interval he sent a letter to his close friend H. O. Blake on September 26 complaining that he was not in a "fit mood" to write, "for I feel and think rather too much like a business man, having some very irksome affairs to attend to these months and years on account of my family."[3]

MEANWHILE, ANOTHER FLOWAGE CONTROVERSY WAS CULMINATING to the north in the Merrimack Valley. On September 28, a group of up to fifty angry men stormed the Lake Village dam at the south end of Lake Winnipesaukee, New Hampshire, with a mind to tear it down. This was the most important dam supplying water to mills of the lower Merrimack River in New Hampshire and Massachusetts. Though the insurgents managed to wrestle a few flash-board planks off the top of the dam, the sheriff showed up in time to prevent further damage. Meanwhile, in nearby Manchester, there was an attempt to blow up the dam of the Amoskeag Company. There a night watchman walking his rounds found ten pounds of gunpowder and an unlit fuse on the face of the dam. Since late August, the dam supervisors there had been personally threatened with violence.[4]

The fuel for the anger in New Hampshire was the same fuel feeding the legal wrangling in Concord. Dam construction had raised the level of Lake Winnipesaukee and flowed the insurgents' lands in an upstream direction. Environmental historian Theodore Steinberg concluded that "their attack on the dam . . . resulted from economic frustration, of lives caught up in the capitalist transformation of the region—an economic shift that left them behind."

Much the same could be said simultaneously about the farmers of Musketa-quid, who had been bucking an irreversible economic trend that would, after the Civil War, make America the richest nation in the world.[5]

Thoreau understood that seething anger. He had felt it as a young man in 1844 when he threatened vigilante justice against the Billerica dam. He had felt it in 1845–1846 at Walden Pond, when he railed against the iron horse of the locomotive, and industrial progress in general. By 1850, how-ever, he had calmed down enough to join that capitalist transformation as a land surveyor. During the mid-1850s he realized that this makeover had, for better and worse, become thoroughly integrated into his landscape. By 1859 his anger had shifted away from engineered structures to political, economic, and cultural issues, especially slavery and financial inequality.

During Concord's flowage controversy, Henry's sympathies clearly lay with the valley farmers who had hired him, rather than the opposing indus-trialists. Nevertheless, observed historian Robert Gross, "the pencil-makers 'John Thoreau & Son'" were "as fully engaged in the market revolution as any other business in Concord." In fact, during the zenith of Thoreau's river years in the late 1850s, he lived in a household "on the fringe of the economic elite," complete with two Irish servants in the house. His surveying business added to the income from the pencil and black-lead business. Based on Thoreau's archive of survey maps, his "eleven jobs in 1859 and thirteen in 1860 are about average for his career." This frequency is confirmed by Henry's self-reporting in his journal.[6]

BY OCTOBER 9, HENRY HAD GONE ELSEWHERE in body and spirit. That evening he gave a speech in Boston for "Theodore Parker's Society" titled "Life without Principle," later published as a well-known essay. His own personal economy as a businessman, surveyor, and legal consultant helped lay the groundwork for the speech.[7]

October 11 saw the third return of the Joint Committee to Concord for local hearings. Judge French opened the case for the petitioners with an ap-peal to theory, explaining that were it not for the dam, meadowland farmers should have witnessed progressive drying rather than progressive flooding: "In the course of time, as cultivation goes on, streams have a tendency to dry up; and mill-streams in all parts of the country have grown more scanty, within the memory of man, some which were formerly capable of driving a large amount of machinery, having quite failed. Here, the case has been precisely the contrary to the ordinary course of nature . . . to say that the Dam does not affect the flow of water above is a transparent absurdity." His argument was right about the upland watersheds but wrong about the Musketaquid.[8]

After his opening argument, Judge French brought to the stand a long parade of witnesses, many of whom were cross-examined by the attorneys for the respondents. He began with old men, whose memories were longest. Oldest was Jonathan Manning, an eighty-four-year-old farmer born two years before the Revolution, in December 1774. This was early enough for him to have helped build the Middlesex Canal dam in 1798, a "zig-zag dam,—leaky and not very high." Under oath, he swore that it "was higher than the former one," meaning that of the Richardson mill. Next was Nathan Barrett, a sixty-three-year-old lifelong resident of Concord who lived where he could look down on a stretch of the river more than two miles long. Beginning in 1828 (the very year the canal dam was raised and strengthened) there were immediate and permanent changes to the stream, with the water coming up higher and staying higher. Stedman Buttrick, a sixty-three-year-old Concord resident who was an occasional boating companion of Thoreau's, confirmed what Barrett reported, adding that the river was now a foot or more higher than before, and that its behavior seemed independent of the rain.[9]

At this point, the leaders of the affected towns began to testify, beginning with Concord's Simon Brown, chairman of the group of "River-Meadow-owners." His testimony was followed by that of John Simonds, of Bedford, who pegged the date for degradation of the river grass to between 1830 and 1840. Wayland's David Heard followed with a litany of complaints that had been part of his family history for more than a century. All three had played a role in hiring Thoreau, and had therefore received his results.[10]

Now it was the respondents' turn to present their case. In his earlier opening statement on October 11, Mr. Butler had assured the committee that "the whole of this evidence" given by the farmers "was one enormous traditionary blunder," and that he could prove this by putting "upon the stand the fathers and grandfathers of the present proprietors," who would "refute every thing their children had averred." He falsely claimed that the 1828 stone "Dam has not been raised since 1798, one inch," and that the earlier timber-built canal dam of 1798 had not been "appreciably raised" relative to the even earlier and more poorly built dams of former gristmills. These statements flatly contradicted the visual truth that the 1798 timber dam was submerged below the waters of the 1828 stone dam, as well as the statements of the petitioning witnesses. Instead of dealing with the facts, Butler argued that the old farmers and town leaders were "men affected by strong prejudice and strong feeling," as they remembered "scenes of their youth as very different from those of to-day." In other words, he was saying, the good old days might not, in fact, have been so good.[11]

The respondents agreed that the meadows had deteriorated, but they shifted the blame to upstream phenomena. Their first correct argument was that the factory reservoirs above Musketaquid had trapped the natural sediment needed to nourish the meadows, causing their deterioration. They also argued that the water being held each winter in the upstream compensating reservoirs was being passed down on the meadows in amounts "more than treble the amount that formerly passed."[12]

Finally, they made the strictly legal argument that the factory owners, "Messrs. Talbot and Faulkner[,] have, acting on the faith of Massachusetts, pledged by the Legislature and undisturbed by sixty years, planted their property and invested their money at these Mills. They employ hundreds of men and women, in profitable employment. There is no fault alleged against them. Yet, because of some fault not even of their fathers, it is claimed, by the Petitioners, that the Legislature should interfere to deprive them of their property. And this is not for public use, but for the benefit of individual Meadow-owners."[13]

The first witness for the respondents was civil engineer John Avery. He confirmed the accuracy of the survey and the water profiles. Next the respondents swore in Lot Esty, who had been hired to monitor the water levels at the factory dam during the experiments, and Israel Colson, superintendent of the dam, who had also helped with the measurements. Their most illustrious expert witness was James B. Francis, an engineer who for many years had been in charge of the waterworks at Lowell. The attorneys described him as the "first authority on the subject in this country, and of reputation throughout the world." Born in Southleigh, England, he had emigrated with his family to Lowell, where he eventually became chief engineer for a coalition of manufacturing companies controlling the flow of the waters of the Merrimack River to the industrial mills along its powerful course. His book *Lowell Hydraulic Experiments,* published in Boston in 1855, became the gold standard for the hydropower technology propelling America into the future. It was on Thoreau's list of readings for that summer.[14]

Francis interpreted for the Joint Committee the experiment of August 5–8, which Thoreau had monitored independently in Concord. After "drawing the water down at the Dam, 34 inches, and keeping it down for four-and-twenty to thirty hours," this "did not reduce the water at the Fordway, more than 5⅛ inches." Furthermore, he testified, "I am certain that the effect of the obstruction caused by the Dam is less, the further you go upstream" and that "at a point ten or fifteen miles up, I can hardly suppose the effect would be perceptible;—it would be less than three inches, at any rate." Francis also correctly pointed out that the dam was only one of three obstructions, the other two being the Fordway and the upstream bars. What he did not say is critical: that these three obstructions were linked like falling dominos. During strong flows, the dam impeded the hydraulic efficiency of the Fordway, which enhanced the flooding, which raised the bar heights, which impeded drainage from upstream agricultural meadows.[15]

The next witness was Daniel Wilson, supervisor of the Middlesex Canal before and after construction of the great stone dam. He flatly stated that the dam had not been raised in 1828. He claimed that the flashboards for that dam, even when fully raised, were the same height as those of the older, 1798 dam. You, the reader, can contradict this statement by going to North Billerica and having a look, or by using Google Earth to see that the older dam is submerged by the present one, built in 1828. His testimony also documented the remains of failed attempts to drain Musketaquid: "At the Fordway I have noticed rocks thrown out at the sides and center of the River, as if blasted. . . . There were some hundred tons taken out, there. I have used as much as one hundred tons, and there is some yet. They hoisted them up with a crane, I have heard them say. I saw the stones beside the River."[16]

George Baldwin—brother of the civil engineer Loammi Baldwin, whose base map Thoreau used for his scroll map—gave the exact height of the dam as $8^{28}/100$ feet. His testimony was the only one to mention channel hydraulics. After stating that "velocity is an essential element," he followed up with "I have never . . . taken the velocity as an element." He had no idea that Thoreau had done so, and used that element in critical calculations.[17]

BY OCTOBER SIXTEENTH THOREAU HAD BEGUN to pull together the first draft of a massive botany project, published posthumously as *Wild Fruits*. But the most important date that month was October 19. First, the hearings moved into their final phase in the statehouse in Boston, where another appropriation of $300 was needed to keep the process going. The second event was the news of John Brown's raid on the federal arsenal at Harpers Ferry, Virginia, and Brown's subsequent capture. Hence October 19 marks a major transition for Thoreau's journal, as his thoughts shifted dramatically away from river science and natural history and toward political morality and abolitionism. Between news of Brown's arrest on October 19 and his execution by hanging on December 2, Henry made no observations relevant to his river project. By then, the freeze-up was well under way. And with the exception of a skating reconnaissance to map and measure river openings on December 23, he put off further investigations until the following spring.

On October 30 Henry gave his lecture "A Plea for Capt. Brown," which asked for a stay of execution for John Brown. On November 1 he gave that same talk in Boston as a substitute for Frederick Douglass, the famed orator and abolitionist who had fled to Canada to escape arrest for his alleged role in Brown's raid. Brown's execution took place on December 2. On that day, Henry Thoreau, Simon Brown, John Shepherd Keyes, Bronson Alcott, Charles Bowers, and the Reverend Edmund Sears held a commemoration of Brown's life in Concord. The first three of these men were heavily involved in the meadowlands controversy and likely would have discussed recent developments.

THE HEARINGS DRAGGED ON THROUGH THE DARK days of the winter solstice. Through it all, the petitioners and respondents remained diametrically and adamantly opposed. Hearings resumed in the statehouse in Boston on October 19 and continued through December 22. To meet their end-of-year deadline, the committee's secretary, Representative Wrightington, complained that they "sat two or three weeks, their sessions continuing the whole day, and sometimes, during the evening also; making, in all, over thirty days in which they had been in constant and close examination of testimony and argument in this case."[18]

Finally on December 22, "after listening to very full and able arguments from Judge Mellen and Judge Abbott, on the one side and on the other," the Joint Committee officially found for the petitioners: "That their lands are materially injured by flooding, and that this injury is far greater than in former years, would seem to have been proved beyond the shadow of a doubt." They identified the dam as one of four causes for increased wetness, the others being the release of summer water from reservoirs, the bridges that narrowed the streams at their points of crossing, and the obstruction caused by bars, weeds, and soil in the bed of the river. All true. They also found serious fault with the unrealistic legal requirement that claims for damage must be made within a year of action at the dam.[19]

At no point during the Joint Committee hearings did anyone realize what Henry did: the bars controlling the water level of the meadows during normal summer conditions had been raised during floods made worse by the Billerica dam. The failure to understand this simple connection—one that every student of fluvial geomorphology learns today—was largely responsible for the confusion that kept the case going (a third fiscal appropriation, on December 23, 1859, would be needed to finish the committee's work). On that same day, Henry returned to his pro bono river science. With good skating conditions, he began to map, measure, and characterize the openings in the ice, and explain them with respect to river currents linked to channel form. These thermal observations put the face of winter on his summer measurements of current velocities.[20]

The New Year

On January 1, 1860, Charles Brace came to Concord to speak, bringing with him a copy of Charles Darwin's *Origin of Species,* published only five weeks earlier. This copy came by way of the botanist Asa Gray of Harvard University, America's strongest proponent of Darwin's ideas. That evening Frank Sanborn held a dinner to introduce Brace and the book to his friends Thoreau and Bronson Alcott. Very soon thereafter, Henry avidly read *Origin,* took six pages of notes in a commonplace book, and informed Sanborn of his approval. By January 14 Darwin's main idea of gradual rather than punctuated change began to influence Henry's journal.[21]

Three days after the Darwin dinner, on January 4, the Massachusetts legislature began its three-month-long debate on the report of the Joint Committee, which was officially released on January 28. Apparently this massive, 607-page report (21 pages of front matter, 370 pages of text, 108 pages of appendices, and approximately 100 untallied pages worth of plates, tables, and foldouts) was pulled together in slightly more than one month. Its recom-

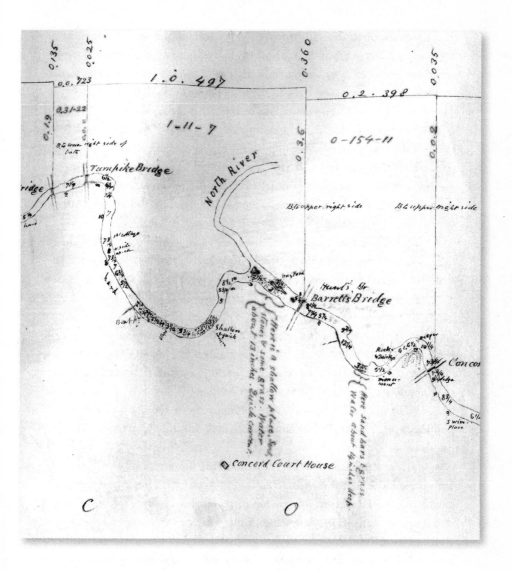

FIGURE 22.   Detail of scroll map. Portion of Thoreau's 1859 scroll map of the
three-river confluence showing his annotations, labels, symbols, and leveling
profiles from the 1833 [of the 1834 map] and 1811 surveys. River survey section
from Survey Document 107a of online exhibit of Thoreau's surveys. Courtesy of the
Concord Free Public Library.

mendations seemed straightforward: to give legal power to the incorporated meadow owners, to grant them funds to clear out the weedy bars on the river, and, most important, to appoint a commission to negotiate with the factory owners to lower their dam back to the level of Osgood's dam of 1710. On January 31 the House of Representatives ordered that these recommendations be referred to a new Joint Special Committee consisting of five members of the House (Messrs. Taylor, of Chicopee; Bishop, of Lenox; Sanborn, of Rockport; Pratt, of Worcester; and Stoughton, of Gill). On February 3, 1860, two members of the Senate were added (Messrs. Bailey and Boynton). By February 24 the report had been concurred by the Senate, the House, and the Committee of Agriculture.

Thoreau was still following the case, as shown by the thirdhand report in his journal entry on February 17, 1860: "Minott says that he hears that Hurd's [sic] testimony in regard to Concord River in the meadow case was that 'it is dammed at both ends & cursed in the middle,' i.e. on account of the damage to the grass there." By February 26 Thoreau had resumed monitoring river stage relative to his original datum at Hoar's steps. One week later, on March 5, he wrote: "The greatest & saddest defect is not credulity, but our habitual forgetfulness that our science is ignorance." Such statements are normally construed to indicate Thoreau's transcendental leanings, but in this case he may be referring to the morass of the legal and political proceedings, failing to take into account the complexity of natural river science.[22]

During the final stages of deliberation on March 8 and 9, an addendum to the original report was printed and presented to the House of Representatives by David Lee Child, of Wayland, the hardest hit of all the towns. His document (which, in the language of that era, was called a memorial) was attached to and circulated with the original petition during its final deliberations. Child claimed that the original documents, as signed by the hundreds of meadowland owners and their town officials, had been materially altered by subsequent committees, and that his memorial corrected those omissions. In it he laid the blame squarely on the improvident legislation incorporating the Middlesex Canal, and asked for complete removal of the dam, a "very feasible reduction of three or four bars, made worse by the Dam," and a "restored right of action against the City of Boston for all damage done by their compensating reservoirs, so called." His document was the first to suggest that the dam made the other bars worse. In short, his "dearest aim" was "to reach and exterminate the harassing and humiliating despotism of the Dam," which he claimed created a "uniform Dead Sea" of the whole valley.[23]

John Shepard Keyes—Thoreau's Harvard classmate, fellow abolitionist, river sympathizer, and a powerful political force in Concord—claimed that during the final month before passage, he played a key role in this legislative victory by getting the charter of the Middlesex Canal Corporation "forfeited" by the state's Supreme Judicial Court. "Then," he added, "I drew the bill for the relief of the meadow owners by taking down the Billerica dam." Quoting his self-aggrandizing autobiography: "I had many hearings before the committees about it," and "after much consultation, a hearing and a view &c &c with lots of lobbying [I] carried it successfully through both houses and saved it from a veto by my influence with [Governor] Banks, to whom I explained it satisfactorily and had the satisfaction of a great triumph with Gov Brown in this vital interest of Concord."[24]

April 4, 1860, was a day of jubilee for the meadowland farmers. After nearly a half-century of lawsuits, they finally won their case. "An Act in Relation to the Flowage of the Meadows on Concord and Sudbury Rivers" authorized the governor to appoint a third commission of three officers to permanently "remove the dam across the Concord River at north Billerica" to a level "thirty-three inches below the top of an iron bolt" that had long since marked the top of the reservoir pool. The act also protected the commission from potential lawsuits associated with the dam lowering.

Project Resumes

The 1860 breakup of the river came late, owing to a particularly brutal, long winter. Though the flowage controversy was now officially over (so it seemed), Henry continued to monitor river stage relative to his datum. It's possible he wanted baseline data to see how the permanent lowering of the dam—scheduled for September 1—would change the long-term hydrology. So on April 25 he reset the datum at his boat place and calibrated it against several other points: "I fix a stake on the west side of the willows at my boat's place, the top of which is at summer level and is about ten and a half inches below the stone wharf there." From that day forward, and with gaps of up to a week or more, he tracked the river level for the next five months through September 27, 1860. By that time, the dam teardown was back in legal limbo, and the river was nearly three feet above the top of the stake, making it difficult to measure.[25]

On May 6 Henry learned something new about the river from his new data set: that high water not only flooded the meadows but also recharged the flanking sandy aquifers along the full length of Musketaquid. This put into storage an enormous volume of water that could later drain back into the channel at a declining rate, ensuring that the river would be kept flowing even during the most pernicious droughts. Meanwhile, the upland swamps would desiccate and harden because they were not similarly recharged by the water transfer: "The swamps are exceedingly dry. On the 13th I walked wherever I wanted in thin shoes in Kalmia Swamp, and to-day I walk through the middle of Beck Stow's. The river meadows are more wet, comparatively."[26]

During his summer monitoring program, Thoreau matched six storm hydrographs (rapid rise and slower fall in stage) with specific precipitation events. For three of these events, the details allowed him to draw important conclusions relevant to the human impact on the watershed. Based on the timing of the flood peaks, he implicated the upstream dams—particularly the compensating reservoirs built by the City of Boston—in delaying and flattening the flood peaks by holding back water during rainstorms and letting it out later. His journal entry for June 10 illustrates the depth of his analytical rigor (and offers some of his worst writing):

> Having ceased raining the eve of the 6th the river reached its highest the eve of the 7th but it had done more than ½ its rising before the rain was over—& it by no means rose steadily afterward—for in the 1st 12 hours (of night) (after) the rain ceased it rose only 1¾₁₆ inches in the next 12 hours (of day) 2¹¹⁄₁₆ in the next 12 hours (of night it rose ¼ inch in the next 12 hrs (of day) it rose 1⅝ in the next 12 hrs (of night) it fell ⅜ ie. in the (first) 2 12 hours of night it rose only 1⁷⁄₁₆ (inch) in the 1st 2½ hours of day it rose 4⁵⁄₁₆ inches. This is the case where the whole rise is 11 or 12 inches at this season—& it—is evidently the effect of the dams above—on one or both streams.[27]

To estimate these effects on the channel near his boat place, Thoreau performed back-of-the-envelope calculations for the flow from the Hopkinton reservoir on the Sudbury River for the period June 13–November 14, and for the flow from the Marlborough reservoir on the Assabet for the period July 14–October 21. These calculations revealed that the average combined summertime release from these reservoirs was a staggering 17,823,180 gallons per day. This translates into 12,377 gallons per minute, which is "about 3½ times the natural supply of the lake," by which he meant Long Pond. Far more water was being sent down into the Concord River in summer than was being diverted away from it by the City of Boston.[28]

As usual, Henry took this line of inquiry beyond what the experts did. He performed an ingenious calculation comparing the discharge from the Sudbury River reservoir (Hopkinton) to its average steady-state river discharge. He found that the reservoir was doubling to tripling the typical late summer streamflow, making the meadows wetter behind their weed-choked bars.[29]

That summer Thoreau learned from direct experience just how wet the Sudbury meadows were relative to the Concord River. With the latter showing "extreme low water . . . you can push over the greater part of the Sudbury meadows in a boat." The only plausible answer for this observation is that the river was dammed up somewhere between these two places. Otherwise, he reasoned, the Sudbury meadows could not be "low" relative "to the river." Though he did not identify any blockages at this time, the data for the 1861 investigation did: the natural sediment dams of the boat-place bar and Robbins Bar.[30]

June 20 closes a gap in Thoreau's stage monitoring record. By June 21 the river had risen seven inches, and it kept rising strongly to crest at nearly sixteen inches on June 23 before a progressive and decelerating fall to a low on July 4. As with his early analysis on June 10, the pattern of the flood hydrograph for this storm indicated the capture of water in upstream reservoirs and aquifers during the rise of the flood, and its erratic release for downstream factory use after the flood.[31]

While waiting for the dam to be torn down that summer, Henry found his main interest shifting toward his botanical manuscripts. By the end of June, concluded Walter Harding, he had gathered together a manuscript "some 575 pages in length called 'Notes on Fruits and Seeds.'" This bulk would be posthumously edited by Bradley Dean and published more than a century later as *Wild Fruits* and *Faith in a Seed*.[32]

On June 29, Nathaniel and Sophia Hawthorne returned to Concord to live. In their honor, Thoreau threw a strawberry party at Emerson's. After a fourteen-year hiatus, their friendship had resumed.

ON JUNE 30, and as an unusual extension of his river project, Henry began an important study of water temperatures in springs, brooks, ponds, and the river. These he compared with temperatures in the air and in soil. This project followed by three years an idea he had in 1857 that "it would be worth the while, methinks, to make a map of the town with all the good springs on it, indicating whether they were cool, perennial, copious, pleasantly located, etc." Apparently the river project of 1859 stimulated him to follow up on this idea, but he didn't have a chance to do it that summer, working instead on fluvial geomorphology, watershed hydrology, channel hydraulics, and sediment transport. Good data for such an investigation must be collected in midsummer, when temperature contrasts reach their widest spread: he had missed his chance the previous year.[33]

Henry's thermal study took about two weeks, and then continued sporadically through August 10. He collected an excellent data set for twenty localities over that twenty-day period. A typical text entry reads disjointedly: "Try the temperature of the springs and pond. At 2.15 P.M. the atmosphere north of house is 83° above zero, and the same afternoon, the water of the Boiling Spring, 45°; our well after pumping, 49°; Brister's Spring, 49°; Walden Pond (at bottom in four feet of water), 71°; river at one rod from shore, 77°; I see that the temperature of the Boiling Spring on the 6th of March, 1846, was also 45°, and I suspect it varies very little throughout the year."[34]

By July 7 he had collected enough data to prepare a table and write a summary report on the spatial and temporal differences in temperature of groundwater springs, brooks, rivers, ponds, and soils, including temperature gradients. This is a robust and highly systematic report. His journal gives his interpretations in a series of tables and cumbersome texts. His salient conclusions are these: The "average temperature of seventeen [springs] is 49½°." He found very little variability in temperature, no more than two degrees. "On the whole, then, where I had expected to find great diversity I find remarkable uniformity" close to the mean annual temperature. His mean temperature for springs is colder than that of today, an interpretation consistent with recent local climate change based on botanical phenology.[35]

Brooks were warmer and more variable than the springs. Cold ones came from "cold, peaty, or else shady swamps," even those that were drained and cultivated. The warmer brooks came from "dry, open uplands." "Brooks are also colder than the river, warming as they enlarge." The summer river was always in the mid-seventies, warmer than the warmest of brooks, usually by about seven degrees. Though the temperatures of the shore and the middle are constantly changing, there was a temperature reversal each day. In the morning the middle of the river was warmest, and in the evening it was the coldest. Cool weather minimized the temperature differences, proving that the heat was solar in origin.[36]

After taking the temperature of springs, brooks, and the river, Thoreau turned his attention to Walden Pond, which put into perspective his observations of pond temperatures from ten years earlier. "So it appears that the bottom of Walden has, in fact, the temperature of a genuine and cold spring, or probably is of the same temperature with the average mean temperature of the earth, and I suspect, the same all the year. . . . Walden, then, must be included among the springs, but it is one which has no outlet,—is a well rather. It reaches down to where the temperature of the earth is unchanging. It is not a superficial pond,—not in the mere skin of the earth. It goes deeper."[37]

This thermal study was the final piece of the puzzle that Thoreau needed to link his waterworld into one coherent package. Essentially, the temperature of springs revealed the timings and pathways of water traveling slowly underground. The deepest springs were the coldest and most stable, having been fed by snowmelt and spring infiltration, and having traveled far enough through aquifers to come very close to the mean annual air temperature. Walden Pond was counted as one of these deep springs. He understood that the route of groundwater everywhere in Concord was down into the earth about a hundred feet or less, and then sideways within aquifers for much longer distances to emerge as cold springs. Such deep springs were not to be confused with warmer springs born at shallower levels above what we now call perched aquifers. He could feel the contact points between groundwater and surface water when wading near the edge of ponds and along the submerged banks of the rivers.

During his groundwater investigation, Thoreau simultaneously monitored the air temperature at corresponding field localities. His thermal datum was the shady side of his Main Street house. From this he learned that rivers were coldest where they were most isolated from the surface air and radiation, which is where they were deepest. He proved this with measurements of the deep bedrock pool at Bittern Cliff on August 21. Brooks have different sources of water, carrying snowmelt in the spring and aquifer drainage in the summer. Starting as cold springs with temperatures in the high 40s and low 50s Fahrenheit, Thoreau showed that they warmed as they enlarged, exchanging progressively more heat with the atmosphere on the way to joining the river. Because the river was merely the sum of large brooks, it was warmer still, and warmest where it was most sluggish. This gave it a chance to maximize solar absorption. Finally, the edges and top of the summer river were warmer than at depth, where continuous seepage from deep and shallow aquifers occurred.

12 | Reversal of Fortune

SEPTEMBER 1, 1860, was the day scheduled for the teardown of the Billerica dam, as specified by state law. "Messrs. Hudson of Lexington, Bellows of Pepperell, and Bigelow of New Bedford" were getting ready to execute the teardown when they were stopped by a court injunction. The factory owners, Talbot and Faulkner, had somehow convinced a judge to halt the process. Concord's J. S. Keyes was outraged by this local strategic move; he believed that the owners of the "river meadows were being cheated out of all they hoped from their bill," which had state and national significance.[1]

"When the day came for these [commissioners] to take down the dam," Keyes wrote in his autobiography, "they were met by a bill in equity asking for an injunction on them upon the pretense that the damages were not secure by the obligation of the state to pay them." This "pretense" typified the whole of the flowage controversy, which had always been more about legal maneuvers by vested interests than about the scientific causes beneath the appearances that Thoreau was investigating. Keyes emphasized that the lateness of the injunction did not allow the petitioners time to "file an answer . . . and have a hearing" that summer, because key members of "the board took that opportunity to go West and be gone." This put off any hope of a hearing until "it was too late that season to do anything more with the dam." Keyes suspected that bribery was involved: "Who paid their expenses of the trip I wish I knew."[2]

Simon Brown, president of the Meadows Association, was also outraged. Editorializing in the *New England Farmer,* he later accused the newly elected General Court of corruption. He wrote that the previous legislature considered the dam "a public calamity, destroying a vast amount of property and spreading desolation and death through one of the most lovely and fertile regions of the State." But "in the meantime fall elections were corrupted by the test question 'Will you pledge yourself to urge and vote for the repeal of the bill directing the dam be taken down?'" After the elections, he described a widely

circulated anonymous pamphlet that was full of "gross misrepresentations."
He alleged that "most of the members" of the General Court "had been visited by the Dam-holders themselves, or their agents." This he considered a
"most shameful and unjustifiable 'lobbying.'"[3]

ON SEPTEMBER 9, eight days after the teardown was supposed to have taken
place, Henry explored the lowermost reach of the Concord River below the
Billerica dam when making a lecture trip to Lowell. It was near here that,
twenty years earlier, he had traveled with his brother en route to the Merrimack in 1839. This reach wasn't part of his river project because it lay downstream of the alluvial valley, where the legal wrangling and scientific investigations were taking place. Nevertheless, Thoreau's descriptions of this lower
reach bring his study of the Concord River all the way to its mouth. He specifically contrasts this lowermost, heavily industrialized reach of the Concord River with the six agricultural reaches on the main stem, where there
wasn't a single dam for twenty-five miles. Thoreau also puts the Billerica dam
in context. Though a single villain to the upstream farmers, it was simply the
uppermost one of a group of four in the same vicinity. For these industrial
people, the entirety of Musketaquid was little more than a colossal millpond.[4]

On September 13 and 14 Henry made the serendipitous discovery that
the channel near his boat place behaved very much like a coastal estuary,
having top and bottom currents flowing in opposite directions within a single
channel. The top current was a coherent mass of warmer Sudbury River water
flowing gently northward over a stronger southerly current of colder Assabet
River water moving up that valley. Here there were two rivers in one place,
moving in opposite directions.[5]

On September 20 Henry gave a speech at the Middlesex Society Cattle
Show in Concord, titled "The Succession of Forest Trees." The day before
and the day after that speech, he was busy monitoring river stage, proving
that his river project overlapped in time with his most important contribution
to proto-ecology. Two weeks later, on October 6, this lecture was published in
Horace Greeley's *New-York Weekly Tribune*. It's a wonderfully argued, albeit
obliquely presented, work that he began four years earlier, and which was
indirectly informed by his meadow investigations. The last of Henry's systematic observations of river stage for the year came on September 27. After con-

templating the persistence of Heywood Meadow on October 22, he combined his observations of plant succession and river history to propose that it takes a "geological change" to create such meadows.[6]

BY NOVEMBER 1860, the temporary injunction to stop the Billerica dam tear-down had been "dissolved" by the Supreme Judicial Court of Massachusetts. The original ruling that "authorized the commissioners to lower the dam thirty three inches" was declared "valid and constitutional." The successful legal argument invoked an "instrumental" conception of the law that had previously long been used to favor industry. Agricultural leaders hailed the decision as a final vindication, and once again the appointed commissioners began their plans to tear the dam down.[7]

And once again the factory owners fought back. Having been overruled by the highest court in the Commonwealth, they changed tactics and sought a political solution: sending out a clarion call to their employees, their families, and the towns they all paid taxes to, making their case for the repeal of the state law. Handbills, miscellaneous propaganda, newspaper advertisements, and the rumor mill became instruments in a campaign of disinformation claiming that the surveyed profiles were inaccurate.[8]

In Thoreau's personal papers is a "Statement to the Public" published in Lowell in 1860, a propaganda pamphlet insisting that the dam be left standing. Cleverly, it asked not for an "unconditional repeal" but for an act "authorizing the appointment of commissioners" to "ascertain by actual experiment the effect produced on the meadows by the drawing down of the river at the dam." Scientifically, this was nothing new. Drawdown experiments based on surveyed river profiles were integral parts of the 1811, 1834, and 1859 studies. Advocates of repeal turned the issue into a debate between jobs and the environment that reads like a harbinger of modern political debates captured by Richard White's essay "Are You an Environmentalist, or Do You Work for a Living?" Specifically, they asked "members of the Legislature of 1861 to consider" keeping the dam for the sake of "the interests of their constituents, the tax-paying citizens of the Commonwealth," and asked them "to act discreetly and justly." Intense lobbying commenced.[9]

By then, Thoreau was too ill to participate, even as a silent third party. December 3, 1860, marks the end of his active, outdoor life. He spent that cold winter afternoon on Nawshawtcut Hill counting tree rings under hypothermic conditions, consequently suffering exposure. Three days earlier, he had an encounter with Bronson Alcott, who was sick with what he called "influenza." And rather than rest and recover from his exposure on the evening of December 3, he let himself be drawn into an argument with Sam Staples—his former jailer—about the morality of John Brown's actions on the first anniversary of his execution. Henry's lingering rancor around issues of morality and slavery is revealed by the following day's journal entry. After beginning with "The first snow, four or five inches, this evening," it moves immediately into "Talk about slavery!" followed by three pages of abolitionist rant. For the next eighteen days there are no journal entries at all, the longest gap in years—a result of the onset of the disease that would eventually claim his life.[10]

Political Solution

One month later, on January 2, 1861, the government of Massachusetts changed leadership. The administration of Republican governor Nathaniel P. Banks was replaced by that of Republican governor John Andrew, who was apparently less sympathetic to the plight of the meadowland farmers. The very next day, on January 3, Thoreau finally admitted to himself just how ill he was, writing that he was "confined to the house by sickness."[11]

When housebound, Thoreau's view of nature became circumscribed by the frame of his sanctum's west-looking windows, which overlooked the Sudbury River. So it's no surprise that on that day he asked himself: "What are the natural features which make a township—handsome?" His answer was "A river—with its waterfalls & meadows—a lake—a hill—a cliff or individual rocks—, a forest—and ancient trees standing singly—Such things are beautiful." In the grammar of this sentence, his river and its watershed, "in the midst of which I was born and dwell," become one and the same: the self-organized, complex, forever-flowing system whose secrets he had explored through years of close acquaintance, record keeping, and quantitative analysis.[12]

On that day he proclaimed that America's rivers should be considered public commons, perhaps because, they had become his de facto commons. "Not only the channel, but one or both banks of every river should be a public highway—The only use of a river is not to float on it." Did this insight about riparian forests influence his final revisions of his essays "Walking" and the

"Wild," which were published posthumously under the former title. When Thoreau wrote: "The west of which I speak is but another name for the Wild," he was looking west over the river. In that essay, he compared his practice of sauntering to a river's practice of meandering, and his favorite direction for walking aligns with an upstream pathway.[13]

January 14 brought Thoreau's first explicit journal reference to Darwinian gradualism. Geologic change, he now believed, involved not sudden re-creations of the world but steady progress according to existing laws, one tiny gradual step at a time. Though this statement was largely correct about Darwinian prehistory, it did not apply to the century of Darwin and Thoreau. During their portion of the Anthropocene, the rate of species extinction and the rate of permanent landscape transformation far exceeded the averages for previous geological epochs. For example, Thoreau's journal chronicles the rapid, wanton extinction of the passenger pigeon, which has since become an icon for thoughtless human planetary impact.[14]

A warm day in February 1861 brought Thoreau out to hear the bluebirds, perhaps for the first time that year. On March 3 he described recent snowfalls and snowmelts. On the eleventh he highlighted his river corridors, rather than the fields and forests, as being the "principal habitat of most of our species." On the twenty-second he returned to the theme of Darwin's natural selection, writing that each species "suggests an immense and wonderful greediness and tenacity of life . . . as if bent on taking entire possession of the globe whenever the climate and soil will permit." This, of course, is exactly what was taking place during his era, with the U.S. population rising 36 percent in a single decade, 1850–1860, most of it in the eastern half of the nation.[15]

On this same day he finally acknowledged that his exposure to the elements the previous December was the cause of the "bronchitis" he was suffering from. One week later, on March 30, he walked down to the river to gauge its height against "Smith's second post." A week after that, on April 6 and 7, he made his final observations of the river: "Am surprised to find the river fallen some nine inches notwithstanding the melted snow." Keeping track of river stage was a habit he couldn't break. Left behind on the bank, he noted, was a residue like "gossamer."[16]

Thoreau's last day as a boatman coincided with a day of victory for the industrialists. On April 9 he described "worm piles in grass at Clamshell," a place that was far easier for someone in his weakened condition to access by boat than by foot. On that same day, the Act to Suspend "An Act in Relation to the Flowage of the Meadows on Concord and Sudbury Rivers" cleared the desk of Governor Andrew. It stipulated that the previous law was suspended until at least May 1, 1862. It also appointed an official commission to conduct a new round of experiments at the Billerica dam, the fourth since the controversy began. The act mandated that the commission consist of "three suitable and competent persons, two of whom shall be civil engineers, experienced in the management and operation of water." The results of those experiments were to be reported to the governor by January 1, 1862.[17]

The language of this legislative act sealed the river's fate. Asking a majority of politically connected engineers allied with the hydropower industry to decide whether a perfectly good hydropower dam should be torn down is like asking oil executives to abandon a profitable petroleum reserve. Or like asking the fox to guard the henhouse. Would the decision have turned out differently if the law had mandated a committee of three agriculturalists experienced with soils? Or three naturalists experienced with meadows?

Whether knowingly or not, this legislation stacked the deck to ensure that the industrial lobby got the cards it wanted. Chairing the special committee was Charles S. Storrow (1809–1904), arguably the most prominent engineer in the region after James B. Francis, the respondents' chief consultant. Storrow, a Harvard graduate, learned his waterworks from the French School, attending the Ecole Nationale des Ponts et Chaussées in Paris. The second engineer on the committee was Joel Herbert Shedd (1834–1915), a politically connected professional colleague. When appointed, he was prominent in the American Society of Civil Engineers, formed only seven years earlier. The third member of the commission was the attorney Daniel Wells Alvord (1816–1871), a former state senator from Greenfield, a graduate of Cambridge School of Law (related to Harvard Law School), and an appointed district attorney for western Massachusetts.[18]

All three had conflicts of interest. The chairman of the commission, Storrow, was a wealthy textile-mill capitalist in the same business as the respondents, Talbot and Faulkner. After returning from France, he joined with six other hydropower industrialists to incorporate the Essex Company in 1845. Nearly from scratch they built a "New City on the Merrimack," later named Lawrence after one of the wealthiest stockholders. That city's economy was utterly tied to the Great Stone Dam across the Merrimack River. With a height of 35 feet and a length of 900 feet, it was the largest dam in the world at the time, and is now recognized as one of the great engineering feats of the

nineteenth century. Storrow was its chief project engineer. With his appointment to the 1861 engineering committee, Governor Andrew was asking the
builder of the world's largest factory dam to decide whether the lowly factory
dam at Billerica was too big. The second engineer, Herbert Shedd, was linked
to the respondents' consultant James Francis through their co-leadership
of the American Society of Civil Engineers. The society's transactions also
document Shedd's previous paid experience with the Assabet and Sudbury
Rivers and his work for the City of Boston's water supply at Lake Cochituate,
which required construction of the compensating reservoirs. Additionally,
someone with the surname of Shedd was paid to supervise the field operations
of the 1861 experiments, suggesting nepotism. The other appointed member,
Daniel Alvord, was something of a third wheel to the mandated majority of
engineers. As a state district attorney, pleasing the governor was part of his
job description. Oddly, he, rather than Storrow, is credited as first author of
the commission's *Report of Experiments and Observations on the Concord and
Sudbury Rivers in the Year 1861*. Was the writing mainly his?[19]

THE AMERICAN CIVIL WAR BEGAN ON APRIL 12, 1861, with the shelling of Fort
Sumter by Confederate troops. At the time, John Shepard Keyes wrote that
"the Legislature . . . was in a bad way" and that Gov. Andrew "was very
busy equipping the militia with overcoats and corresponding and advising in
every direction." Despite his concerted lobbying efforts to gain "some relief
for the river meadow case," Keyes saw his attempts go nowhere. Understand-
ably, a three-year-old investigation of wet meadows fell rather low on the list
of government priorities.

Meanwhile, a gravely ill and house-bound Thoreau was getting worse.
The young Horace Mann Jr. was visiting him and giving him a series of presents that only a naturalist could enjoy: the "contents of a stake-driver's [the
heron-like American bittern] stomach or crop," and the same for a "black duck."
Mann was also shooting birds to bring: a hermit thrush, a pigeon hawk, "two
small pewees . . . a white-throat sparrow, Wilson's thrush, and myrtle-bird."
On May 11, an unidentified boy gifted Henry with a salamander.[20]

After five months of confinement in his home, and under pressure from
his doctor, Henry made plans for a trip to Minnesota, believing that the dry
continental air might cure what was now clearly a case of consumption. His

new young friend Mann agreed to be his traveling companion. Starting and ending their trip by railroad, they left Concord on May 11, returning on July 9. When out West, Henry was especially impressed with the size of the mighty Mississippi below Minneapolis and by the tortuous meanders of the Minnesota River, which put anything he'd seen on the Sudbury River to shame. Ironically, Henry's last good river observations were of the most distant rivers he encountered during his life. In a long letter to Franklin Sanborn dated June 25 from Red Wing, Minnesota, he writes of the "grandeur and beauty" of the Mississippi, its lake-speckled headwaters in the northern pine country, and the breadth of its entrenched valley south of Minneapolis. Its tributary, the Minnesota River, was so meandering that it was without "a straight reach a mile in length as far as we went,—generally you could not see a quarter mile of water, & the boat was steadily turning this way or that."[21]

The hoped-for cure of dry western air didn't work. Thoreau returned to Concord sicker than before and utterly exhausted. Once again confined to the house, he was no longer able to sojourn on foot. A carriage ride in July, offered by Simon Brown, helped him see the country without the exertion of boating or walking. In August Thoreau made a final visit to Daniel Rickertson in New Bedford, during which he had a photo taken, the famous ambrotype by E. S. Dunshee, reproduced in this book. After he returned to Concord, death was just a matter of time. By October, he was too weak to walk outdoors, so he took a horse and wagon instead, courtesy of neighbor E. R. Hoar. Sister Sophia became his steady companion, nurse, and editor.[22]

Engineering Commission

As Thoreau's life energy ebbed, the hydraulic experiments required by the act passed in 1861 were being carried out on the Concord River. The full record is available as *Report of Experiments and Observations on the Concord and Sudbury Rivers, in the Year 1861*. As with the 1860 report, the investigations were confined to Musketaquid, even though the Assabet played a key role. The Engineering Commission Report was completed on the deadline of January 1, 1862, and published later that year in Boston by William White, printer to the state.[23]

This was a very unscientific study. Despite the mind-boggling estimate of 35,000 measurements, no statistical analysis was carried out. No novel explicit hypothesis about the dam was tested. Rather, the drawdown experiments were clones of the 1859 and 1834 experiments, but with better data. The two novel parts of this study were mainly asides. The first was a subprogram to monitor standing water on the meadows with respect to water in the channels. The water there was mostly perched. The second novelty was monitoring the before-and-after effects of clearing weeds on local river stage.

For the 1861 drawdown experiments at the dam, thirty new stations were established between the Billerica dam and the Beaver Hole Meadows above Wayland. The exact elevation of each station was referenced to the level line surveyed by John Avery during July 1859, which he had previously discussed with Thoreau. All of these stations were referenced to the iron bolt marking the historical level of the dam pool, and assigned an arbitrary datum elevation of 10.00 feet. Variations in river stage at these thirty stations were then monitored on a preplanned schedule by forty-six paid observers, many of whom were paid boarders in the homes of the petitioners. Local residents cooperated by loaning the observers boats and helping with the weed-clearing equipment. At twelve of these stations the water levels in the adjacent meadows were tracked as well, using dug holes and wooden stakes. After weeks of study, the data were presented mainly as summary tables, from which John Avery drew a series of water profiles for different dates and different conditions. The total costs were $7,501 for the data collection and $4,700 to compensate the Talbots for lost power during times when the dam was drawn down. In 2017 dollars, the cost approached $1 million.

The final engineering report drew three duplicitous conclusions. First was that the dam "is not the only, nor the chief cause of the wet state of the meadows above." Second was that lowering the water level at the dam did have an impact on the river as far as Robbin's Bar, but it was trivial: "A drop of 16½ inches at low flow will, in the ordinary summer conditions of the river, reduce the level 8 inches at the fordway, 6¼ inches at Barretts Bar, and disappear above Robbins Bar. Third was that keeping the river free of weeds "will . . . reduce [the] level at Wayland and Sudbury about 6 inches, but not [give] substantial relief."[24]

The first conclusion conflates prehistory with history. Because the meadows have been there for thousands of years, the generally "wet state" must be prehistoric in age: the natural hydraulic damming of Musketaquid's bedrock outlet at the Fordway. But this wasn't what the court case was about. It was about whether the Billerica dam made the "wet state" wetter, which their data proved to be the case. Their new survey showed that the top of

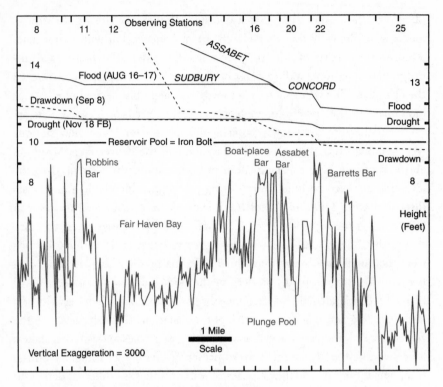

FIGURE 23. River profiles. Surveyed profiles for Thoreau's three rivers by John Avery in 1861. Fathomed bed and named bars in gray. Black lines show height of water surface for three rivers during three stage conditions. The slopes of the water surface profile relative to each other and to the bed constrain hydraulic interpretations. See text for explanation. Base is Alvord, Storrow, and Shedd 1861.

the Billerica dam was higher than the bottom of the river for all twenty-five miles upstream to Saxonville. Additionally, their own data showed that the outlet's hydraulic efficiency was being compromised by the dam pool. This first conclusion is also unscientific. The commissioners studied only two of multiple causes for wetness (the dam and weeds), yet concluded that the effects of these factors were less than unstudied others. Finally, the first conclusion is blatantly refuted by evidence they ignored. The November 18 river profile shows that under normal dry summer conditions and with the flashboards raised (Drought [Nov 18 FB] in Figure 23), the river profile was flattened like a lake for the full length of the valley. In other words, the high water of normal factory operating conditions greatly reduced the current over places where, under other conditions, the flow would have been swifter.[25]

Their second conclusion ignores a guiding axiom of fluvial process: that a river's base flow discharge takes place within channels set by high-flow conditions, which were in part controlled by the Fordway. This is clearly shown by comparing the flood profile for August 16–17 (during which water was flowing well above the dam crest) with the profile for drier conditions on September 8. It shows that the three most important sediment bars (Barrett's, boat-place, and Robbins) were submerged and being shaped by the August flood that the Commission ignored. When the flood subsided, each bar became a sediment dam that created an upstream lakelike reach that kept the water high on the meadows.[26]

Their third conclusion is only partially correct. Yes, clearing weeds does increase drainage and therefore lower the water level. But claiming this as an improvement was based on the erroneous assumption that the bars impeding flow were "natural and original," meaning they were primary bedrock and glacial obstructions, rather than those formed by the accumulation of "sand, gravel, or other sediment in the bed of the river." This statement is clearly true for the Fordway, where there were few weeds, but not for the other two bars, where the weeds throttled back the flow. By 1861, the conditions of Barrett's Bar and Robbins Bar were definitely not "natural and original." Rather, they were the result of historic sediment deposition at the highest water levels ever.

THE TEXT OF THE ENGINEERING COMMISSION'S REPORT claims that each commissioner was "furnished with a full record of all the observations, amounting to upwards of 35,000 in all, examined them by himself, and formed his own conclusions without consultation with the others." It also claims that on "all the more important questions. . . . the three commissioners came separately to the same conclusions." If this is true, then how can we explain why an attorney with no expertise in channel flow came to the same conclusions as two gifted hydraulic engineers? And why did all three commissioners independently fail to mention the single most important piece of evidence, the November 18 data set and resulting profile? And why was the low total "fall" never mentioned, when this was the single most important numerical value considered by the previous surveys of 1859, 1834, and

1811? And why did all three commissioners poke fun at those who might be interested in the high-water (August 16–17) river profile when, in fact, it was this flow that created the sediment dams that were limiting to meadowland agriculture?[27]

The unanimity of faulty logic and errors of omission for the 1861 report make me wonder if any of the three commissioners gave more than a passing glance to the mountain of data before deciding the dam should remain as it is, rather than being torn down to a lower level. The big difference between the state-commissioned and state-funded 1859 and 1861 investigations involved the way these opposing decisions were reached. The former was an open legislative process in which five elected appointees from throughout the Commonwealth acted like a cross between a legislative study committee and a legal jury. For ten months, judges guided them through due process, through various stages of hearings, deliberations, and reporting. This open process was later upheld by the state Supreme Judicial Court. In contrast, the 1861 process was a generally closed and overwhelmingly political process. A three-person committee dominated by private waterworks engineers with egregious conflicts of interests used unscientific experiments to reach a result that pleased the governor who appointed them. In hindsight, a commission chaired by a textile manufacturer decided to leave a dam standing to support textile manufacturing. Looking back from the twenty-first century, the whole thing appears to be a tactical delay followed by a whitewash.

Two Deaths

On January 1, 1862, Governor Andrew communicated the 1861 report to the General Court with the endorsement that "you will, I think, be found a complete and satisfactory repository of the scientific truth sought through its agency." Not so.[28]

By this time, Henry Thoreau had resigned himself to death, and his sister Sophia was helping him take care of final business. His friends from Worcester, Harrison Blake and Theophilus Brown, skated between Framingham and Concord to pay their final respects. Between February 20 and 28, Henry and publisher J. T. Fields agreed to bring the out-of-print *Walden* back into print, a deal for 250 copies. On April 12, Fields visited Henry and Sophia in person to finalize the arrangements, and to buy the unsold copies of *A Week* to bring out as a new edition.[29]

On that same day, Judge French made his final plea on behalf of the petitioners. By then the meadowlanders were desperate enough to switch tactics, agreeing that removal of the Billerica dam would not bring significant relief to the upstream meadows. Instead, French asked for help in cutting out the occluding bars. On the following day, March 13, Judge Abbott summarized his final arguments on behalf of the respondents. After presenting the same timeworn arguments, he warned the legislature not to make this case a precedent, lest the farmers of every other valley in Massachusetts follow suit. This, he strenuously argued, was a test case for what we now call environmental law.

On March 21, with a foreboding of death, Thoreau wrote to a literary correspondent named Myron Benton: "If I were to live, I should have much to report on Natural History generally." This and all subsequent letters would be written in Sophia's hand. Six days later, on March 27, a newly appointed Joint Committee of the Legislature finally recommended the repeal that Governor Andrew had asked for.[30] By mid-April, Henry's death was imminent. At his request, his bed was moved downstairs from his isolated third-floor sanctum to the front parlor on the main level. Never again would he see the sun sparkling over the river at his boat place. Shortly before his downstairs move, Henry was visited by his old friend Ralph Waldo Emerson, who wanted to pay his respects and say goodbye. Emerson recalled the water view from the western window, and he remembered the way Henry spoke of the river as if it were another old friend he'd known for a lifetime. Emerson later recorded Thoreau's final autobiographical comparison between his life and that of his river: "Henry praised to me the manners of an old, established, calm, well-behaved river, as distinguished from those of a new river. A new river is a torrent, an old one slow and steadily supplied." The younger Thoreau was a boatman with wild, agitated, idealistic ideas. The older Thoreau was a boatman whose mature flow of thoughts was slower and steadier. Emerson finished his entry with his friend's final words on the topic: "What happens in any part of an old river relates to what befal[l]s in every other part of it." Metaphorically, the moments of our lives and the movements of our rivers finally all flow together.[31]

Concord commemorated the eighty-seventh anniversary of the "shot heard round the world" on April 19, 1862, with a send-off at the railroad station. Forty-five Union soldiers, all volunteers from Concord, boarded the train to fight America's first and only civil war. For some, their caskets returned on the same rails.

On April 25, the General Court of Massachusetts passed An Act to Repeal "An Act in Relation to the Flowage of the Meadows on Concord and Sudbury Rivers." Its language was brief and to the point: the previous ruling to tear the dam down "is hereby repealed," effective immediately. Its passage, eleven days before Henry's death, was also the death knell for meadowland farming as a way of life. Simon Brown, writing anonymously as the editor of the *New England Farmer,* called this finality an "unjust and wicked oppression upon an unoffending and long-suffering people."[32]

On the morning of Henry's death, May 6, 1862, Sophia was reading to her brother from *A Week*. Anticipating its ending, the glorious, fifty-mile-long sail home, Henry uttered the last coherent sentence of his life: "Now comes good sailing." A triple entendre, it describes an actual memory, quotes his first book, and invokes a metaphor for passing from this world to the next. For the boatman at death's door, sailing with a "smacking wind directly aft" would have been a happy thought to end on.[33]

Conclusion

APPROXIMATELY ONE YEAR BEFORE HE DIED, Thoreau had a good laugh about the practice of law in general and water law in particular. "I hear that Judge Minott of Haverhill once told a client, by way of warning, that two millers who owned mills on the same stream went to law about a dam, and at the end of the lawsuit one lawyer owned one mill and the other the other." This black humor from the April 11, 1861, entry in his journal nicely summarized the final result of the flowage controversy. When the gavel came down in the General Court at Boston on April 25, 1862, the result was a big fat zero, except for three years' worth of gainful employment for the attorneys on opposite sides, and for those within the legislature.[1]

After more than 1,100 days of meetings, hearings, experiments, and writing sessions coordinated by half a dozen government-funded committees and commissions, the final result looped back to where it all started. The last of four legislative acts repealed the first. First came the act to appoint a Joint Committee to study the situation (April 1, 1859). Next, based on that study, came the act to tear down the Billerica dam (September 1, 1860). After that came the act to suspend the teardown and study the matter once again (April 9, 1861). Finally came the act to repeal the initial act, which brought everything back to the beginning (April 25, 1862). All of this time and money, especially during preparations for Civil War, could have been saved by asking one local genius to weigh in.

Of course, Thoreau would not have rendered the Solomon-like judgment that the law so craves. Rather, after eighteen months of river investigations, he had become convinced that the entire watershed of Musketaquid above its natural outlet was behaving as one big coherent system within which humans were pervasive and ubiquitous players.

UNDENIABLY, MUSKETAQUID WAS DAMMED AT THE BOTTOM. But was it *damned* there as well? As a youthful rebel in his mid-twenties, Thoreau certainly thought so, because he threatened to bring it down with a crow-bar to liberate the fishes and help his farmer-neighbors. After studying the whole system carefully in 1859–1860, however, the forty-five-year-old natural scientist knew that placing sole blame on that lower dam was misguided. Not once after the summer of 1846 have I found anything in Thoreau's commonplace books, journals, or correspondence explicitly indicating he wanted the dam torn down.[2]

Thoreau had proof beyond reasonable doubt that raising the dam made the meadows wetter all the way up to Wayland. He wasn't fooled by the vagaries of the drawdown experiments. Instead, he used local history and common logic. The Fordway had been properly named in the seventeenth century because it was then a ford: an easy place to cross the river except in flood. On July 31, 1859, he linked this historical fact to a modern drawdown experiment, using an exclamation point for emphasis: "I am told that formerly you could sometimes walk through there without its going over your boots! & indeed that that was done quite lately within a few weeks when some repairs were made on the dam." This qualitative comparison is more convincing to me than the avalanche of quantitative stage data the Engineering Commission poured down on the problem. The river was fordable either before the dam was built or when the water was returned to that previous level. For a bedrock streambed, these are identical hydraulic situations. "That it is not used as a fordway of late years," Thoreau concluded, was the single "best evidence that the water is deeper there than formerly." The "water lines" on the rocks there do not lie.[3]

Quantitative proof came from the drawdown experiments he monitored. Dropping the dam pool 16½ inches brought the water level at the Fordway down by half that amount, proving bottom-up control by the dam. When the dam pool was allowed to come back up, the result was a much gentler gradient and a stagnant channel at depth, both of which reduced the hydraulic efficiency of Musketaquid's natural outlet. This meant higher and longer-lasting floods, which meant higher gravel bars at the T-junction. When the floods subsided, the higher boat-place bar turned the lower Sudbury River into a higher lake. Thoreau recognized the historic ecological change there in his lifetime: "The Hibiscus & White maple do not occur on the main stream for a long distance above the mouth of the Assabet maybe 10 miles." This proved that the extra wetness there was recent, and therefore part of the downstream story of human impact. The 1861 monitoring data would later prove him correct.[4]

Undeniably, the Assabet and Sudbury Rivers were dammed at the top by the dams of the compensating reservoirs built by the City of Boston, and those of countless factory mills. But were they *damned* by that? The meadowland farmers thought so, calling the summer water releases from the compensating reservoirs "water avalanches," and comparing their violence to an alligator in the water. Thoreau's high-resolution hydrologic data from two separate lines of inquiry during 1860 showed that the compensating reservoirs strengthened and raised the summer flows, causing more frequent floods and wetter meadows.[5]

He also bemoaned the wholescale deforestation of the uplands, which had been taking place for centuries. By the winter of 1852, he was alarmed that "they are cutting down our woods more seriously than ever. . . . Thank God, they cannot cut down the clouds!" By 1857 he was watching razed land being burned in all directions, and predicted a completely denuded landscape in the near future: "The smokes from a dozen clearings far and wide, from a portion of the earth thirty miles or more in diameter, . . . woods burned up from year to year. . . . The smokes will become rarer and thinner year by year, till I shall detect only a mere feathery film and there is no more brush to be burned." By 1859, forest cover had reached its minimum: "There is scarcely a wood of sufficient size and density left now for an owl to haunt in, and if I hear one hoot I may be sure where he is."[6]

This deforestation was drying the uplands and enhancing the storm flooding. Sprout-land and trodden pastures were sending stronger pulses of snowmelt runoff to the main valley, leaving less to infiltrate down to aquifers to sustain the brooks between storms. Thoreau saw this as a crime against upland landscapes: "These little brooks have their history. They once turned sawmills. They even used their influence to destroy the primitive [forests] which grew on their banks, and now, for their reward, the sun is let in to dry them up and narrow their channels. Their crime rebounds against themselves."[7]

The upland environmental makeover made Thoreau's rivers more volatile than before. This was particularly true for late summer subtropical storms. A heavy summer rainstorm that previously would have been buffered by forest foliage, soaked up by humus, and allowed to infiltrate down to the aquifer instead now drained away as surface and shallow subsurface runoff to nearby brooks. When these brooks combined forces, and when they entered sediment-filled downstream channels, they overflowed, strengthening the

summer freshets that ruined the hay. And with the forest gone, far less moisture was being sent back up to the atmosphere via evapotranspiration during
the summer. This kept Musketaquid's summer stage higher than during the
Holocene.

With higher storm discharges came excess sediment, especially from
tillage landscapes. Thoreau wrote: "The ground being bare where corn was
cultivated last year, I see that the sandy soil has been washed far down the hill
for its whole length by the recent rains combined with the melting snow. . . .
This, plainly, is one reason why the brows of such hills are commonly so
barren. They lose much more than they gain annually." When washed into
his three rivers, this sand gave them material to work into beaches on the
edge of enlarged channels: "various parallel shore lines, with stones arranged
more or less in rows along them—thus forming a regular beach of 4 or 5 rods
length."[8]

Additionally, the upland landscape was being drenched by livestock manure, whether dropped on pastures by the animals themselves or spread on
grain fields as fertilizer. Thoreau began to notice that the green scum called
duckweed, so characteristic of barnyards, was now showing up in river sloughs.
The use of imported chemical fertilizers, particularly guano, was increasing as
well. Swamps that once sopped up nutrients now became a source. Thus, when
it rained, a flush of nutrients worked its way downstream into a sluggish river
where the only thing limiting aquatic growth was the amount of phosphorous
and nitrogen. Ironically, the fertilization of English hay on the uplands stimulated the growth of aquatic plants in the shallows, clogging the river with
weeds and backing up water.

Undeniably, the watershed was changed in the middle by countless
transformations within the subwatersheds of all three of Thoreau's rivers. But
was it *cursed?* Certainly the owners of the Billerica dam thought so, blaming
extra meadow wetness mainly on the loss of sedimentary nutrient, or "slime,"
needed to keep the meadows from decomposing and descending. For centuries harvests had been sustainable because the smallest particles of river
sediment were deposited on the meadows each time they were covered by
water. The dams, however, trapped the annual dose of organic sediment in
their mill ponds. With less fertilization, hay cutting became unsustainable.
Thereafter, vegetative growth was being maintained by the consumption of
muck soils. This translated into a loss of volume, which translated into a loss of
elevation, which translated into an increase in wetness. Finally, the earlier aer-

obic bacterial metabolic pathways had added biomass to the meadows. When submerged, the switch to anaerobic conditions consumed the muck, which contributed to slow subsidence and vented what the complainers called "mephitic gasses," mainly hydrogen sulfide.

Meanwhile, the farmers on those starving meadows were mismanaging them with too much ditching, burning, and mining. Muck was mined to spread on the upland fields in lieu of animal manure. These excavations created artificial ponds that held more water longer, thereby keeping meadows wetter. As they slowly gave up on the alluvial meadows, farmers were increasingly drawn to upland swamps, which could be managed like cranberry bogs to grow hay. Thoreau described one such "improvement": "Here was an extensive swamp, level of course as a floor, which first had been cut, then ditched broadly, then burnt over; then the surface paved off, stumps and all, in great slices; then these piled up every six feet, three or four feet high, like countless larger muskrat-cabins, to dry; then fire put to them; and so the soil was tamed." Drainage of such wetlands reduced their ability to store water, intensifying the flood wave and therefore causing water to back up higher at the Fordway.[9]

Industrialists above the meadows were reallocating the budgets of water, sediment, and nutrient with "eighty-two mill-ponds or reservoirs" run by "some fifty manufacturers." All sent surges of water down to Musketaquid in summer. Some of the dams were in good shape; others were not and could fail catastrophically. Barrett's dam was leaky to the point of comedy: "At the dam, am amused with the various curves of jets of water which leak through at different heights. According to the pressure. . . . The dam leaked in a hundred places between and under the planks." In the upstream direction, the effects of such dams were also easy to see. They created "a permanent freshet. . . . You have only to dam up a running stream to give it the aspect of a dead stream. . . . Some speculator comes and dams up the stream below, and low! the water stands over all meadows, making impassable morasses and dead trees for fish hawks,—a wild, stagnant, fenny country." When the reservoirs were full, this increase in saturated ground produced more storm runoff and a wetter Musketaquid.[10]

Bridges on the main stem created a different problem. Each, he showed, became a local hydraulic dam that, in effect, converted a smooth longitudinal profile into a series of terraced steps. The one-arch bridge on the Assabet, for example, raised the river by at least two feet during floods. This new stair-step morphology was, in turn, amplified during spring breakup when the ice floes choked the channels at bridges and when uplifted tufts of meadow got stuck between the abutments. The threat of both of these potential obstacles—floes and tufts—was amplified by meadow farming practices. With respect to meadow wetness, farmers had become their own worst enemies.

STEPPING BACK FROM THE LEGAL AND ECONOMIC HISTORY of the flowage controversy reveals a surprising story. From the point of view of Thoreau as a boatman, his three-river Anthropocene landscape was an improvement over its early Puritan condition. He never said this explicitly, but it's pretty easy to connect the dots between the changes that he knew were taking place and how he liked to spend his time. When he wrote that "all nature begins to work with new impetuosity on Monday," he was not being critical. The holding back of water on Sundays and its releases on Monday morning was similar in concept to the scheduled releases of stored water from modern dams to improve whitewater kayaking and to restore the natural habitat of rivers, most notably on the Colorado River. This catch-and-release of stream water enlivened Thoreau's daily boating experiences by providing pulses of flow.[11]

It did bother him that these scheduled releases forced his fellow river creatures to observe the "christian sabbath" even as he was trying to avoid it. When he wrote that the Assabet River was being "emasculated & demoralized" by hydropower development, his main lament was that Christianity had penetrated so deeply into nature that "the very fishes feel the influence (or want of influence) of man's religion." The Sunday river showed gravel bars that the weekday river did not.[12]

The more denuded Concord's upland landscape became, the more the riparian forests bordering his three rivers became unfenced sanctuaries for wildlife. These corridors became wooded commons being enhanced by sawdust mulch, extra mineral sediment, and higher levees consisting of uplifted meadow tufts. The higher the water got from dams and bars, the less the gallery riparian landscape was used for agriculture, and the more Thoreau had it to himself. Nine years before his death, he wrote, "in relation to the river, I find my natural rights least infringed on. It is an extensive 'common' still left." From the boatman's view, such riparian woods are visually magnified by a factor of four: two banks and two reflections.[13]

As a boatman, Henry was allied with the downstream industrialists, rather than meadow haymakers. The latter loved times of low water because the hay could be easily scythed, the wagons didn't sink into the mire, the stacks could rest on dry ground, and the cattle could be turned into the meadows to graze the aftermath. Henry and the factory owners preferred the higher and steadier flows of late summer courtesy of the compensating reservoirs, which is why the reservoirs were built in the first place: to prevent drought. For the industrialists, the streams of water and cash were closely linked. For the boatman, drought was a bummer. It brought death to the meadows, dust to the lily pads, an oily sheen to the water, and a mephitic stink to the air. That's when Thoreau compared his rivers to the accursed rivers of antiquity. In short, upstream water releases from reservoirs great and small during summer improved both his life as a naturalist and the cash flow of factory workers.

With more flood runoff coming down from above because of forest clearing and channel clogging, and with the river's natural outlet being throttled back by the Billerica dam, Thoreau's opportunities for blue-water sailing on the Great Meadow and the fjordlike Carlisle Reach were much improved. Virtually every drenching storm lasting more than a day caused back-flooding. On these occasions Henry would launch his boat, paddle down the dredged channel of Barrett's Bar, and sail that inland sea until it drained away. This joy was more frequent, more intense, and more long-lasting because the chain of meadow lakes had become deeper and better connected. In a triumph of wildness over civilization, Thoreau would sometimes sail joyously over the flooded road network.

The lower Assabet offered a completely different kind of boating joy. Stronger river flows enlarged his paddling habitat and enhanced its aquarium-like beauty. This clear-water stream assumed a bed composed of yellow sand and pebbles. Strong bank undercutting offered sylvan shade at the Leaning Hemlocks. Shoals at scour holes such as the Assabet Bath were rippled with golden sands derived from erosion. Downstream from Egg Rock, this pulse of sediment lengthened, widened, and raised the sediment bars in what Thoreau called the "rapid reach" of the Concord River. This invigorated the paddling experience between the slackwater reaches above and below. At the time scale of days, its flood waves were more dramatic, and therefore more exciting. At the scale of hours, the surface went up and down like clockwork based on the work schedule of upstream mills. This made it more interesting.

The Hill-Lake reach of the Sudbury River—Clamshell Bank, Nut Meadow Brook, Conantum, and Fairhaven Bay—was rendered more lakelike because its outlet over the boat-place bar had been raised. In consequence, channel obstructions were submerged more deeply. A pulse of ecological change affected the shorelines. Meadows became broader and more variable, attracting Thoreau's botanical interests. Its principal tributary, Nut Meadow Brook, carried more water during storms, making it more interesting than it had been in earlier times. It became Henry's best laboratory for observing hydraulic processes, and the place where he first understood the physics of the meandering process.

With 90 percent or more of the forest cut down, the land was everywhere exposed to stronger winds. The prevailing regional winds from the northwest and southwest were less dampened by the fluttering of leaves and the swaying of branches. Replacing the bending, groaning trees of the original forest was the cropped grass of pastures and the fields of tillage lands. The open Anthropocene uplands extracted far less wind energy, leaving stronger residual winds to blow across more open water to fill the sails of Thoreau's boat. Additionally, the locally generated winds of spring and fall were stronger and more regular, owing to the greater thermal contrast between cleared land and open water. These local winds rippled his blue waters with a greater sense of motion and increased the number of days on which he could sail.

Another impact of regional deforestation was more water for Musketaquid relative to upland soils. The removal of mature trees—many of which can consume hundreds of gallons of water per day through evapotranspiration—left more of the summer water budget available for streamflow. From this effect alone, Thoreau's three rivers would have been higher and their currents stronger during his Anthropocene than during the Holocene of his Native American predecessors.[14]

Finally, the deliberate engineered conversion of Long Pond into Lake Cochituate by the City of Boston gave Thoreau one of the most inspiring nautical scenes of his life. For two straight days he gushed about its "glorious sandy banks far and near, caving and sliding,—far sandy slopes," on the edge of a "vast and stretching loch on which he might sail." It was like having a smaller version of Cape Cod Bay within the Concord Valley. As with the sandy banks and bars of the Assabet River (created by fluvial adjustments to its one-arch bridge), the shorelines of Lake Cochituate were experiencing a wild and dramatic pulse of shoreline erosion, in this case precipitated by raising the dam at its outlet to divert the flow to Boston. Beyond the improvement in nautical scenery, this also gave Thoreau a new relation to the soil: "When I see her sands exposed, thrown up from beneath the surface, it touches me inwardly, it reminds me of my origin."[15]

CONCORD'S MAKEOVER BEGAN WHEN ENGLISH COLONIALS arrived to create a religiously based communal village on the banks of this lush river meadow. Brian Donahue proved that theirs was a largely sustainable agrarian culture. Yet within a decade they were trying to reengineer their "Great River" landscape by ditching its bedrock outlet. When that failed for lack of technology, they explored the possibility of diverting floodwaters eastward through the meadows of Thoreau's birthplace, over the watershed divide, and into the Shawsheen Valley. Their pioneering efforts inaugurated more than two centuries of human "improvements" driven by an exponentially rising American population and improved technologies. Maximum development—agricultural land conversions and industrial hydropower development—occurred in Thoreau's generation. By the time of his death, virtually every square inch of the land had been transformed in some way: rock to quarry, river to canal, brook to millstream, wood to fuel, land to pasture, and field to cropland. The entire valley had become a network of dams, canals, turnpikes, railroads, factories, town streets, and country roads.

John Barry, in his history of American river engineering, describes the cultural zeitgeist of Thoreau's century: "This was the century of iron and steel, certainty and progress, and the belief that physical laws as solid and rigid as iron and steel governed nature, possibly even man's nature, and that man had only to discover these laws to truly rule the world. It was the century of Euclidean geometry, linear logic, magnificent accomplishments, and brilliant mechanics. It was the century of the engineer."[16]

Local monuments to that century were the Billerica dam, the Middlesex Canal, the Fitchburg Railroad, and the Union Turnpike. Each was a battle won in a war against nature launched by General George Washington's Revolutionary Army. Indeed, on June 16, 1775, Washington appointed Colonel Richard Gridley of Massachusetts to be the army's first chief engineer. At the time, the principal concern was military fortification. This remained the case in 1802, when Thomas Jefferson stationed a corps of engineers at

West Point, creating the U.S. Army Corps of Engineers. In 1824, when Thoreau was a happy barefoot boy, the Corps broadened its scope into civil affairs, beginning a war against America's rivers that intensified as the Corps became the default federal agency for river management. Initially the technical knowhow came largely from the French School of hydraulic engineering. Not until 1835, midway through Thoreau's college years, did Rensselaer Polytechnic Institute in New York grant the first civilian American engineering degree. "Even up to the 1850s," wrote Corps historian John Chambers, "nearly all of the engineers—military and civilian—had received their scientific education at West Point." During peacetime, rivers dominated their mission. Charles Ellet Jr.'s *The Mississippi and Ohio Rivers* (1853) and Andrew Atkinson Humphreys's *Report upon the Physics and Hydraulics of the Mississippi River* (1867) summarized river civil engineering in their times. Everything was about control.[17]

Since the dawn of antiquity, rivers, even more so than coasts, have been the primary battlefields in this war against nature. Geologist Ellen Wohl provides a nice summary in her 2004 book *Disconnected Rivers*. The U.S. Geological Survey, acting through the U.S. National Research Council, summarized the situation in 2007: "Deforestation, industrialization, urbanization, floodplain cultivation, dams and levee construction, and channelization have altered dramatically natural flow regimes." Thoreau, by the end of his life, had encountered all of these disruptions and disconnections in his own field studies.[18]

His pioneering river science of 1859–1860 did not position humans as masters and commanders of their watersheds, as did the engineers of his day. Instead, he saw human actions as hopelessly entangled with natural ones. Attempts to "civilize" the landscape became perturbations, from which the wildness of nature emerged automatically elsewhere. The clearest example of his thinking involves Barrett's Bar, where sawdust and sand were interstratified. By definition, that sawdust is granular organic sediment of complex origin. To make it requires soil creation, forest ecology, deforestation, the building of sawmills, and the cutting of wood into the stuff of Main Street life: its houses, furniture, vehicles, containers, and so on. By definition, the mineral sand on Barrett's Bar is also granular sediment. To make it requires crystalline rock, glacial crushing, diluvial deposition, human exposure, a pulse of bank erosion, and sand redeposition in high water backed up by the Billerica dam. Thus, even something as seemingly simple as Barrett's Bar is an integrated manifestation of geology, ecology, hydrology, and technology at the scales of the whole watershed and of millions of years. And the location of the bar was largely set by the upriver edge of the flood pool backed up by a fordway being controlled by a dam. Thoreau's insight was not that his rivers were being impacted by human activities, and not even that his rivers were the result of human activities, but that appearances and causes were so entangled that legal finger-pointing was pointless.

The most comic example of this entanglement was Thoreau's realization that the Concord River religiously "keeps the Sabbath." This makes the timing of river flows an artifact of human cultural practices dating back to the late Bronze Age of the ancient Hebrews on the other side of the world. Thus it was that Old World religion became part of New World hydrology.

The most complex and important example of entanglement was Henry's understanding of how raising the Billerica dam at the bottom of the valley created the worst flowage in Wayland, at its top. Had the attorneys and engineers of Thoreau's time known what he knew, they likely would have settled out of court. Raising the pool behind the Billerica dam enhanced flooding, which enhanced the boat-place bar, which became the sediment dam for a higher lake in the lower Sudbury River, into which poured Pantry Brook, also known as Gulf Brook. It responded quickly and powerfully to local storm events, in part because it had been heavily disturbed by two centuries of bog-iron mining, charcoal production, and land conversions. Flashy local flows from Pantry Brook brought excess sediment to the upstream edge of the higher lake. This sediment lodged at the first sharp downstream bend, creating or raising Robbins Bar, which the Sudbury River was too weak to cut down during low flow. Thus Robbins Bar became the low-flow sediment dam for the Sudbury meadows, where the complaints had always been loudest.

HENRY PROBABLY STOOD ALONE IN KNOWING HOW RIVERS WORKED as natural systems, at least in the United States. During his era, the gulf between engineering and natural science was much larger than today. The civil engineers were not being trained to think scientifically, and the natural scientists were not paying much attention to river behavior. This was certainly true for the geologists of Thoreau's era. Aside from the obvious fact that rivers flushed sediment to the sea, geology didn't properly claim river behavior as a subject until well after

Thoreau's river project. As late as 1871 the nation's most influential textbook, Edward Hitchcock's thirty-second edition of *Elementary Geology*, gave almost no attention to river channels. At this time, Thoreau's pioneering river work lay unpublished and unappreciated in his attic garret.[19]

George Merrill's monumental summary *The First Hundred Years of American Geology*, published in 1904 by the Smithsonian Institution, says almost nothing about river channel form or sediment transport beyond a brief mention of the Mississippi River and the reinterpretation that windblown dust in the lower Mississippi Valley wasn't a fluvial deposit. At this moment in the history of American science, Thoreau's journal had not yet been published, and his works had been largely forgotten. Yet within his archive was evidence that Thoreau, working as a lone genius, had correctly interpreted many of the key ideas of fluvial geomorphology a half century before the subject was invented. Much of this has been missed by Thoreau scholars, who have instead fixated on Thoreau's well-known criticisms of institutional science for being falsely objective and detached from meaning. What scholars typically fail to see is that Thoreau was addicted to the puzzle-solving practice of science as a routine part of his daily sojourning life.[20]

Well into America's Gilded Age, during the last quarter of the nineteenth century, federal interest in geology, which accelerated after the California gold rush of the late 1840s, was focused almost entirely on mineral resources. Not until 1877 did geologist Grove K. Gilbert, a pioneering geomorphologist, demonstrate for American scientists that river channels were landforms in equilibrium with their environments, adjusting their shapes, sizes, gradients, and banks in response to new conditions. Not until 1888 did the U.S. Geological Survey begin a program of river science and stream gauging under the direction of John Wesley Powell, and at that stage the effort was mostly about water supply. Not until the 1890s did the physical geographer William Morris Davis provide a theoretical framework for long-term change in river behavior. Not until 1914 did fluvial geology become a robust experimental science, with Gilbert's field evidence coherently integrated into a proper theory.[21]

TO ME, THE MOST ASTONISHING THING about Thoreau's river project is his recognition that the natural world he fell deeply in love with was no longer wild in a traditional sense, but was even wilder with respect to the dynamism and unpredictability of landscape process. Human inventions and interventions were propagating down every available energy gradient toward his future and our present. One of those changes was the return of pastures, woodlots, and cultivated fields to "wild" second-growth forest on upland watersheds. The regional hydrology responded with a decline in net annual runoff because the vigorously growing forests were taking more and more of the annual water budget. This left less water in the summer Concord River during Thoreau's twentieth century canonization than during his lifetime. Within the last half century, however, the trend for New England is an increase in annual streamflow owing to greenhouse warming. The regrowth of forest since Thoreau's death also changed the local meteorology. With an increase in turbulence associated with the second-growth canopy, and with more evapotranspiration, the regional winds are now lighter and the air more humid. Both of these changes have decreased the quality of boating relative to that of Thoreau's lifetime. He was indeed born in the "nick of time" with respect to these atmospheric variables.[22]

None of this is to suggest that we should blithely accept the great harm we are doing to the planet today. Indeed, we are creating severe and irreversible problems, particularly to low-lying nations being submerged by rising sea levels, specialized habitats such as the Arctic pack ice of the polar bear being eliminated, and coral reefs being blanched in acidifying oceans. Thoreau was rightly angered by the wanton extinction that preceded him, by his own complicity in land development, by the gluttony of biofuel consumption causing the last of his woodlots to be cut down, and by the rise of coal consumption in his day. Thoreau specifically criticized Ralph Waldo Emerson's household for burning as much as "twenty-five cords of wood and fourteen (?) tons of coal" each year during the late 1850s.[23]

Thoreau's point is that the world we live in was and is being changed in ways and on time scales that we don't even recognize. Going forward, he would want us to accept and deal with those changes without romanticizing or sanitizing the past; to take advantage of the good and mitigate the bad. We must learn to adjust to each situation on a case-by-case basis with clearly stated positions and goals. For example, a higher dam at Billerica was good for Thoreau as a boatman but bad for his haymaking friends. The mature Thoreau would likely agree that the conservation of land, water, habitat, and species must be done with objective local goals in mind, rather than some dreamy notion of how good things used to be.

Looking back from the late 1850s, Thoreau decided that his three rivers had changed little in his lifetime. At first glance, this seems to contradict the dramatic pulse of change he was experiencing. But keep in mind that he didn't build his first boat until 1833, when he was only sixteen years old. By then, the big impacts of 1827–1828 caused by raising the Billerica dam and mis-designing the Union Turnpike Bridge were already going strong. Already, his rivers were flowing with greater "impetuosity" than before, thanks to the landscape makeover he was part and parcel of.

Epilogue

AT THE UPSTREAM END OF HENRY'S SCROLL MAP, the Sudbury River wanders off into meadows so vast that he didn't bother to draw them. At the downstream end of his map, he drew a fair likeness of the Billerica dam. To see the dam today, all you have to do is drive to North Billerica, park on the street, walk over to the small community park, and look out over the rails. There you will see the Concord River arching over the dam's crest to foam in the turbulence below. Its water is a faint amber color in fall, tinged green in peak summer, and nearly clear in winter. Behind the dam is a flat sheet of water extending farther than you can see.

That dam was the main battleground for a statewide fight between meadowland farmers and textile industrialists. Before that local war was over, the Union and Confederate armies were engaged in a national civil war over secession and slavery. Armed militia from formerly competing towns of Middlesex County were now riding the rails together en route to battle. The South had become a common enemy that united valley residents. The flowage controversy dropped off the list of state priorities. The wet, degraded meadows above the Billerica Dam were given up for lost. Left largely unnoticed, they continued to deteriorate for the next half century. Less hay was cut each year, and then only in the driest places. Meanwhile, the channels were being heavily polluted by factory chemical waste and sewage effluent until cleanup began in the late 1960s.[1]

Was the 1860 court decision to tear down the Billerica dam the right one? Yes, with respect to the legal process. Yes, based on the evidence weighed by the Joint Committee. And yes, as a prelude to the dam controversies of the twentieth century. By 1908, Thoreau's writings were inspiring John Muir and the Sierra Club to fight the proposed damming of California's Hetch Hetchy Valley to supply water for San Francisco. Because a national park was involved (Yosemite), this was America's first truly national debate on environmental policy. Muir lost his fight, the park was compromised, and the O'Shaughnessy dam remains today. By the early 1960s Thoreau's and Muir's examples were inspiring David Brower, Edward Abbey, and thousands of conservationists to prevent the Colorado River from being dammed as well. They lost their fight, and the Glen Canyon Dam remains today. For another decade or so, U.S. federal agencies, especially the Bureau of Reclamation, the U.S. Army Corps of Engineers, and the Department of Defense, kept up the momentum for damming and disrupting rivers.[2]

But in the present century, the momentum has shifted toward river emancipation—even when the generation of hydroelectric power is a clear gain. Hundreds of rivers throughout the United States are being unshackled, thanks to river science. Fish ladders around the Billerica dam are being seriously considered for the first time in centuries. Henry Thoreau should be credited for his pioneering river studies and his prophetic leadership for river liberation. And the Concord River should be cited as the place where it all began in earnest.

Was the 1862 political decision to leave the dam in place the right one? No, because the Engineering Commission Report of 1861 was unscientific and ethically compromised. Let that report be a cautionary tale about the fusion between money and politics during the rise of America's Gilded Age.

Did everything turn out okay? Yes, based on a lucky stroke of twentieth-century American history. During the early 1900s, recreational sport hunting became wildly popular, thanks in part to the example of President Teddy Roosevelt. This was especially true for waterfowl such as those that Thoreau watched from his boat on the river: geese, ducks, herons, bitterns, ospreys, egrets, and even bald eagles. Throughout the nation in the early twentieth century, wetlands previously deemed of little value were being reclaimed as wildlife habitat for conservation purposes, if only because legions of waterfowl hunters wanted something to shoot at. This cultural shift led to a steady increase in federal legislation protecting wetland habitat, such as the Migratory Bird Treaty Act of 1918. In 1921, Congress introduced a bill to establish a refuge system for migratory birds. Rejected four times, it finally passed as the Migratory Bird Conservation Act of 1929. Its authority created the National Wildlife Refuge System we know today.[3]

Locally, Concord resident Samuel Hoar (1887–1952), a duck-hunting devotee, was part of this shift in the conservation zeitgeist. Recognizing the habitat value of the degraded swampy marsh of the Great Meadow, he purchased a large parcel in 1928. Working in a completely opposite direction from that of his ancestors, he built a series of earthen dikes and dams on the meadow to permanently flood it, in order to tempt waterfowl to land. By 1944 the National Wildlife Refuge System had become a respected and growing federal program whose interests closely aligned with those of Samuel Hoar. So that year he donated 250 acres of his private refuge to inaugurate what is now the Concord Unit of the Great Meadows National Wildlife Refuge, now managed by the U.S. Fish and Wildlife Service. Its outline broadly mirrors the meadows shown on the northern portion of Henry's scroll map.

To broaden protection along the Atlantic flyway, the United States Fish and Wildlife Service began acquiring other derelict land parcels along Musketaquid, especially during the 1960s. The soaking-wet and long-abandoned meadows of Wayland and Sudbury were added as the Sudbury Unit of the Great Meadows refuge. When combined, the Concord and Sudbury units contain more than 3,800 acres of prime wildlife habitat extending along more than twelve miles of Musketaquid, roughly 85 percent of which is wetland. From north to south, this system spans seven towns—Billerica, Carlisle, Bedford, Concord, Lincoln, Sudbury, and Wayland—all of which once fought very hard to drain the water away. Then, in 2005, the U.S. Army transferred more than 2,200 acres of surplus land along the Assabet River to the U.S. Fish and Wildlife Service. The result was the Assabet River National Wildlife Refuge.

In 1635 the Musketaquid meadows were rich enough to nucleate America's first river town. In 1861 Thoreau prophetically asked that "not only the channel, but one or both banks" of his three rivers be declared a "public highway." His request was ignored and his meadows were given up for lost in 1862. Ironically, this loss set the stage for their national protection in the middle of the twentieth century. Today, the Concord, Sudbury, and Assabet Units of the U.S. National Wildlife Refuge System constitute the largest federally protected patch of wild landscape in the midst of New England's most settled urban corridor. And to keep that wildness available for human use, in 1999 twenty-nine miles of Thoreau's three streams were designated "Wild and Scenic Rivers," in part for their historical and literary associations. Thoreau would be pleased to know that his nineteenth-century vision of river wildness continues to sustain us in the twenty-first century.

Abbreviations

Works by Henry David Thoreau

P1   Journal, Princeton edition. *Journal 1: 1837–1844.* Edited by Elizabeth Witherell et al. Princeton, NJ: Princeton University Press, 1981.

P2   Journal, Princeton edition. *Journal 2: 1842–1848.* Edited by Robert Sattelmeyer. Princeton, NJ: Princeton University Press, 1984.

P3   Journal, Princeton edition. *Journal 3: 1848–1851.* Edited by Robert Sattelmeyer. Princeton, NJ: Princeton University Press, 1990.

PT   Journal online transcripts. Manuscripts: Vols. 28, 29, 30, 31, and 32. Edited by Princeton editorial team. Online at http://thoreau.library.ucsb.edu/writings_journals.html.

TP   Thoreau's papers: Henry David Thoreau Papers 1836–1862. Concord Free Public Library Special Collections, Series II. Nos. 107a, 107b, 107c, and 107d. Vault 35A, Box 1, Folders 5 and 6. Scroll Map (107a) and Statistics of Bridges (107b) indexed under "Thoreau Surveys." Correspondence, miscellaneous notes, pamphlets, maps, draft statistics of bridges, and newspaper clippings, in folders.

WA   *Walden, or Life in the Woods: A Fully Annotated Edition.* Edited by Jeffrey Cramer. New Haven, CT: Yale University Press, 2004, originally Boston: Ticknor & Fields, 1854.

WK   *A Week on the Concord and Merrimack Rivers.* Edited by Thomas Blanding. Orleans, MA: Parnassus Imprints, Inc., 1987, reprinted from Boston: James Monroe & Co., 1849.

Reports of the State of Massachusetts

ECR   Daniel W. Alvord, Charles S. Storrow, and Herbert J. Shedd [commissioners appointed under Chap. 154, Acts of 1861]. *Report of Experiments and Observations on the Concord and Sudbury Rivers in the Year 1861*. Boston: William White, Printer to the State, 1862.

JCR   Massachusetts Joint Special Committee. *Report of the Joint Special Committee upon the Subject of the Flowage of Meadows on Concord and Sudbury Rivers January 28, 1860*. Boston: William White, Printer to the State, 1860.

Thoreau Archive

CFPL Concord Free Public Library, Special Collections, Concord MA.

Notes

1. Citations to quotes and key ideas are bundled in sequence at the end of paragraphs. Each is preceded by a keyword or phrase, separated by a colon. Most are to Thoreau's *Journal*. For the duration of Thoreau's river project (March 17, 1859–September 27, 1861), these citations are to the unpublished, online transcripts of the journal prepared by the Princeton editorial team, abbreviated PT, by date and page number. Citations for his "river years" (May 12, 1850–June 7, 1861) are, for consistency's sake, from the 1906 Houghton edition using the date only. Prior to May 12, 1850, I cite the published Princeton editions by date, volume, and page number, using P1, P2, and P3 as abbreviations for volumes 1, 2, and 3, respectively. Other frequently used works by Thoreau are abbreviated WK for *A Week on the Concord and Merrimack Rivers*, WA for *Walden, or Life in the Woods*, and TP for Thoreau's Papers in the Concord Free Public Library (CFPL). Two large reports by the State of Massachusetts are abbreviated JCR, for the Report of the Special Joint Legislative Committee (1860), and ECR, for the Report of the Engineering Commission (1862).

2. Hereafter the "scroll map." A high-resolution digital copy of the map is available at www.concordlibrary.org/scollect/Thoreau_surveys/Thoreau_surveys .htm.

3. Thoreau's Map 107a bears two complete titles from two previous surveys, neither of which is Thoreau's. The map, erroneously called his "river survey," is archived under the title of its 1834 base map.

4. Women: David Wood, curator at the Concord Museum, recalls two rooms, estimated to be nine by twelve feet each, with windows on each side of the house (personal communication, Apr. 13, 2016). Cotton: It was about 600 feet from the river, comparable in distance between his house at Walden Pond and the main part of the lake. Willows: Henry's authorship is also proven by his distinctive handwriting, the presence of draft map sketches within his private papers, hundreds of links to his then-private journal, the use of the possessive personal pronoun "my" on the map's notes, and the fact that his sister donated it.

5. Both ends: Feb. 17, 1860, PT, 5.

6. Engineer: H. F. Walling, *Map of the Town of Concord, Middlesex County Mass* (Boston: No. 31 Washington St., 1852). Oversee: Robert Richardson, "Thoreau and Science," in *American Literature and Science,* ed. Robert J. Scholnick (Lexington: University Press of Kentucky, 1992), 12. Laura Dassow Walls, *Seeing New Worlds* (Madison: University of Wisconsin Press, 1995), 247. Henry was appointed to Harvard's Committee for the Examination in Natural History. Self-described: Dec. 2, 1852.

7. *Vitesse de régime:* Pierre Louis Georges Du Buat, *Principes d'hydraulique et de pyrodynamique: vérifiés par un grand nombre d'expériences faites par ordre du gouvernement,* 3 vols. (Paris: F. Didot, 1816). Eternity: Ralph W. Emerson, *The Letters of Ralph Waldo Emerson,* vol. 8, *1845–1859,* ed. Eleanor M. Tilton (New York: Columbia University Press, 1991), 622.

8. Impetuosity: July 20, 1859, PT, 165.

Introduction

1. The gardening was assisted by an African American man named John Garrison Jr. Franklin Sanborn, *Hawthorne and His Friends: Reminiscence and Tribute* (Cedar Rapids, IA: Torch Press, 1908), 20. Muskmelon: Nathaniel Hawthorne, *Passages from the American Notebooks, Volume 2* (Chapel Hill, NC: Project Gutenberg EBook, 2005), 110.

2. Steed: Hawthorne, *Passages from the American Notebooks,* 111–112.

3. Backyard: Officially, these are the Sudbury, Assabet, and Concord Rivers, though the term "Concord River" is often used to describe them collectively. The term "Concord River Valley" is generally synonymous with the collective watershed. The term "Musketaquid," refers to the flat lowland river valley of the Sudbury and Concord Rivers that is bisected by the town of Concord.

4. Unfenced: May 30, 1852.

5. Successful: *Walden* was only marginally successful in Thoreau's lifetime. Deliberately: WA, 88. Forest land: Apr. 15, 1852.

6. Cover painting: David Foster, *Thoreau's Country* (Cambridge, MA: Harvard University Press, 1999).

7. Clear water: July 2, 1858. Woods: Robert Richardson, *Henry David Thoreau: A Life of the Mind* (Berkeley: University of California Press, 1986), 63. Landscape: William Ellery Channing, *Thoreau the Poet Naturalist with Memorial Verses, New Edition,* ed. F. B. Sanborn (Boston: Goodspeed, 1902), 22. Being: Alfred Tauber, *Henry David Thoreau and the Moral Agency of Knowing* (Berkeley: University of California Press, 2001), 64.

8. Fluvial: July 10, 1852. Rivular: Apr. 6, 1853. Willows: Kenneth Grahame, *The Wind in the Willows* (New York: Charles Scribner, 1908), 11.

9. Essay: Henry David Thoreau, "Walking," *Atlantic Monthly* 9 (1862): 657–674. Smoking: Aug. 30, 1853. Sea-room: Aug. 24, 1852. Current: June 11, 1855. Oars: Aug. 31, 1856. Seafaring: Edward W. Emerson, *Thoreau as Remembered by a Young Friend* (Boston: Houghton Mifflin, 1917), 2.

10. Solitude: After July 29, 1850, P3, 102. Purpose: Channing, *Poet Naturalist*, 22.

11. Unpublished poem from the original manuscript of *A Week* in the Berg Collection of the New York Public Library, cited in Tauber, *Moral Agency*, 64.

12. Died: Ralph W. Emerson, "Thoreau," in *Excursions*, ed. Leo Marx (Boston: Ticknor and Fields, 1863), 18.

13. Chanticleer: June 2, 1853. Namesake: WA, 324. Baker Farm: WA, 194–201. Dwell: Mar. 12, 53. A-fishing: Mar. 25, 1842, P1, 390

14. Nature reborn: Aug. 17, 1851. Every rain: Ibid. Train of thought: Aug. 14, 1854, and Oct. 12, 1851. Proper currents: Dec. 12, 1851. My thought: July 2, 1858. Blood: Feb. 27, 1860, PT 163. Blue arteries: Ibid.

15. Failure: Lawrence Buell, *The Environmental Imagination* (Cambridge, MA: Harvard University Press, 1995), 146, labels it an "utter commercial failure." Doggerel: Emerson, *Friend*, 94.

16. Literary work: Sharon Cameron, *Writing Nature: Henry Thoreau's Journal* (New York: Oxford University Press, 1985). Muskrats: Channing, *Poet Naturalist*, 14.

17. Geometric: John Demos, *Circles and Lines* (Cambridge, MA: Harvard University Press, 2004). Chronologic: At maximum, the Walden epoch is from 1844–1854.

18. Looks west: Mar. 31, 1852. Lake: distances were estimated from Thoreau's graphic scales for the *Walden* bathymetric map and the river scroll map. Pillow's edge: July 22, 1851. Northeast: Ibid. Reflected: Nov. 7, 1858

19. Boat place: located on the scroll map as "boat-pl." Convenient: Feb. 13, 1859. Knows it: Aug. 6, 1858.

20. Retirement: Franklin Sanborn, *Henry D. Thoreau* (Boston: Houghton Mifflin, 1882; reprint, New York, Chelsea House, 1980), x. Picturesque: Ibid.

21. Lake of the Woods: Aug. 5, 1851.

22. Scenery: Dec. 2, 1852. The Sudbury River refers to the main line of drainage both above and below the inlet to Musketaquid. The Concord River refers to the main line above and below the outlet of Musketaquid at the Fordway.

23. Pearly: Aug. 11, 1859, PT, 231. Millers: Though technically freshwater mussels, they were called clams. Anthropocene: Officially, the Anthropocene is a geological epoch newly proposed in P. J. Crutzen and E. F. Stoermer, "The Anthropocene," *Global Change Newsletter* 41 (2000): 17–18. It's now being debated by the International Code of Stratigraphic Nomenclature, mainly on the unresolved issue of its lower boundary, in Stanley C. Finney and L. E. Edwards, "The 'Anthropocene' Epoch: Scientific Decision or Political Statement?, *GSA Today* 26 (2016): 4–10. Culturally, the term has become an important part of the intellectual zeitgeist. See Jedediah Purdy, *After Nature: A Politics for the Anthropocene* (Cambridge, MA: Harvard University Press, 2014). Geologists are increasingly adopting the word, despite its unofficial status. See L. E. Edwards, "What Is the Anthropocene?," *Eos* 97 (2015): 6–7 and W. F. Ruddiman, E. C. Ellis, J. O. Kaplan, and D. Fuller, "Defining the Epoch We Live In," *Science* 348 (2015): 38–39. This book uses the Anthropocene as an unofficial global geological epoch (lower case) associated with the earliest trace fossils in strata associated with European settlement (1636) and continuing in the present.

24. Iron pot: Aug. 11, 1859, PT, 231. At once: Ibid., 232.

25. Discharges: James H. Patric and E. M. Gould, "Shifting Land Use and the Effects on River Flow in Massachusetts," *Journal of the American Water Works Association,* 1976, 41–45. Decades of experimental forestry at Hubbard Brook, New Hampshire, provide the best scientific constraint on the role of New England forest cover and watershed hydrology. See www.hubbardbrook.org /overview/siteadmin.shtml (accessed August 2016). Reduced evapotranspiration from the watershed increased bottomland wetness.

26. Shore: Aug. 11, 1859, PT, 231. Forenoon: Ibid.

27. Egotism: Apr. 2, 1852. Man-worship: Dec. 2, 1852. Earth earthy: Sept. 22, 1854. Thought: Michael Rawson, *Eden on the Charles: The Making of Boston* (Cambridge, MA: Harvard University Press, 2010), 17.

28. Wilderness: William Cronon, "The Trouble with Wilderness," in *Uncommon Ground: Rethinking the Human Place in Nature,* ed. William Cronon (New York: W. W. Norton, 1996), 80–81.

29. It is vain: Aug. 30, 1856. Other authors: Ralph W. Emerson, *Nature* (Boston: James Munroe, 1838). George P. Marsh, *Man and Nature or, Physical Geography as Modified by Human Action* (New York: Charles Scribner, 1864). Leo Marx, *The Machine in the Garden: Technology and the Pastoral Ideal in America* (New York: Oxford University Press, 1964). Richard White, *The Organic Machine: The Remaking of the Columbia River* (New York: Hill and Wang, 1995). David Nye, *America as Second Creation: Technology and Narratives of New Beginnings* (Cambridge, MA: MIT Press, 2003). Richard Judd, *Second Nature: An Environmental History of New England* (Amherst: Univer-

sity of Massachusetts Press, 2014). Diana Muir, *Reflections in Bullough's Pond: Economy and Ecosystem in New England* (Hanover, NH: University Press of New England, 2000).

30. Serious issue: Bill McKibben, *The End of Nature* (New York: Random House, 1989). Long run: Stephen J. Gould, *Bully for Brontosaurus* (New York: Oxford University Press, 2000), 18.

31. Faunas: The official North American Faunal Age names and dates are Racholabrean (>14,000 years BP), Santarosean (14,000–400 years BP), and Saintaugustinean (<400 years BP, or <1636 AD regionally). Nobler: Mar. 23, 1856. Races: Dec. 29, 1853.

32. David Henderson, "American Wilderness Philosophy," *Internet Encyclopedia of Philosophy,* http://www.iep.utm.edu/am-wild/(accessed Nov. 10, 2016). M. Mitchell Waldrop, *Complexity: The Emerging Science at the Edge of Order and Chaos* (New York: Touchstone, 1992).

33. Wallace Stegner, *Crossing to Safety* (New York: Modern Library, 2002), 190.

34. Turbulent Fall: May 22, 1857.

35. Apocalyptic: Jo Guldi and David Armitage, *The History Manifesto* (New York: Cambridge University Press, 2015), 62. Judd, *Second Nature.* Uncanny: Purdy, *After Nature,* 230–231. Appearances: Mar. 23, 1853.

36. *Mosses:* Nathaniel Hawthorne, *Mosses from an Old Manse* (New York: Wiley and Putnam, 1846), 26–27. Victorian: George B. Bartlett, *The Concord Guide Book,* ill. L. B. Pumphrey and Robert Lewis (Boston: D. Lothrop, 1880). Hemlocks: Apr. 1, 1852.

37. Side-hill: April 1, 1852. Slide: Aug. 22, 1853.

38. Ear: Mar. 2, 1859, PT, 205. Pewee: May 7, 1852.

39. Raccoons: Judd, *Second Nature,* 119. Relative to Concord, Thoreau found the true wilderness of the deep Maine Woods to be impoverished of animal life.

40. Heavens, etc.: Aug. 8, 1858.

41. Unctuous: June 30, 1852. Fluvial walks: July 10, 1852.

42. Acres: Boston Journal, "The Meadow Lands of the Concord River Valley," *New England Farmer* 11 (1859): 76–78. Middlesex: Line from famous poem, "Paul Revere's Ride," by Henry Wadsworth Longfellow. Fight: Robert G. Gross, *The Minutemen and Their World* (New York: Hill and Wang, 1976) describes the social context of the "Concord Fight."

43. Society: Morton Horwitz, *The Transformation of American Law, 1780–1860* (Cambridge, MA: Harvard University Press, 1977), 253. Urban industrial: John

Cumbler, *Reasonable Use: The People, Environment, and the State, 1770–1930* (Oxford: Oxford University Press, 2001), 4. Greatest number: Jonathan Prude, *The Coming of Industrial Order: Town and Factory Life in Rural Massachusetts, 1810–1860*, 2nd ed. (New York: Cambridge University Press, 1999), 238. Replacing: Cumbler, *Reasonable Use*, 67. Products: Gordon Eaton, "Down to Earth: A Historical Look at Government-Sponsored Geology," in *The Earth around Us: Maintaining a Livable Planet,* ed. Jill S. Schneiderman (New York: W. H. Freeman, 2000), 86.

44. Report: JCR. (Online document available from the Massachusetts State Library.) The page count is mine, a sum of front matter, text, appendices, and pages needed to cover the many plates.

45. Oath: June 24, 1959, PT, 111. Private papers: TP.

46. JCR, 19.

47. Josiah Abbott, *Argument of Hon. Josiah G. Abbott on Behalf of the Petitioners before the Joint Special Committee* (Boston: Wright and Potter, 1862), 12, 25, 12, respectively.

48. Spindles: Theodore Steinberg, *Nature Incorporated: Industrialization and the Waters of New England* (Cambridge, UK: Cambridge University Press, 1991), 103. Waste water: WK, 102. This debate and Thoreau's project were enmeshed with events leading up to the American Civil War, notably news of John Brown's raid on Harpers Ferry reaching Concord on October 19, 1859, South Carolina's secession from the Union on December 20, 1860, and the attack on Fort Sumter on April 12, 1861.

49. Steam-tug: July 23, 1856.

50. Commissioners: June 24, 1859, PT, 114.

51. Otters: Feb. 4, 1855.

52. Walter Harding, *The Days of Henry Thoreau* (New York: Dover, 1982 [1965]), 439. Thoreau, "The Succession of Forest Trees," in *The Succession of Forest Trees and Wild Apples by Henry D. Thoreau, with a Biographical Sketch by Ralph Waldo Emerson* (Boston: Houghton, Mifflin, 1865), 33, 52.

53. Scientific value: Bradford Torrey and Francis Allen, "Introduction," in Thoreau, *The Journal of Henry David Thoreau in Fourteen Volumes Bound as Two* (New York: Dover, 1962), 219.

54. Wrong: Chura, *Land Surveyor,* 156–157 describes the excisions as "ill advised." Canonization: Buell, *Imagination,* 345.

55. P1, Henry David Thoreau, *Journal,* vol. 1, *1837–1844,* ed. Elizabeth Witherell, William. L. Howarth, Robert Sattelmeyer, and Thomas Blanding (Princeton, NJ:

Princeton University Press, 1981), 587. Harding, *Days,* 411. Buell, *Imagination.* Richardson, *Thoreau.* Laura Dassow Walls, *Seeing New Worlds.* Laura Dassow Walls, *Material Faith: Henry David Thoreau on Science* (Boston: Houghton Mifflin, 1999).

56. Sara Luria, "Thoreau's Geopoetics," in *GeoHumanities: Art, History, Text at the Edge of Place,* ed. M. Dear, J. Ketchum, S. Luria, and D. Richardson (London: Routledge, 2011), 126–138. Patrick Chura, *Thoreau the Land Surveyor* (Gainesville: University Press of Florida, 2010). Daegan R. Miller, "Part One. River Tree: Henry David Thoreau's Countermodern Cartography," in "Witness Tree: Landscape and Dissent in the Nineteenth-Century United States," Ph.D. dissertation, Cornell University, 2013, 22–87. Buell, *Imagination,* 278–279.

57. The manuscript journals are PDF files within Manuscripts: Volumes 28, 29, 30, 31, and 32 being edited by the Princeton team. Online at thoreau.library .ucsb.edu/writings_journals.html. Experience: For a primer on fluvial processes, I suggest John S. Bridge, *Rivers and Floodplains: Forms, Processes, and Sedimentary Record* (Oxford: Blackwell, 2003).

58. Projects: Buell, *Imagination,* 138, describes Thoreau's projects, and this wasn't considered one. Geocentric: Robert M. Thorson, *Walden's Shore: Henry David Thoreau and Nineteenth Century Science* (Cambridge, MA: Harvard University Press, 2015), 8.

59. Eulogy: Emerson, *Thoreau,* 7–33.

60. Beaver: Robert J. Naiman, Carol A. Johnston, and James C. Kelley, "Alteration of North American Sterams by Beaver" *Bioscience* 38 (1988): 753–762. Deforestation: David Foster et al., *Wildlands and Woodlands: A Vision for the New England Landscape* (Cambridge, MA: Harvard University Press, 2010), 5, Figure 1. Transition: May 11, 1860. Climate change: William R. Baron, *Historical Climates of the Northeastern United States: Seventeenth through Nineteenth Centuries,* in *Holocene Human Ecology in Northeastern North America,* ed. George P. Nicholas (New York: Plenum Press, 1988), 29–46. His historical reconstructions for eastern Massachusetts show a steady increase in ambient moisture from the 1830s to the 1860s from "drier" to "wet," and the percentage of fair days per summer falling from 75.7 to 47.4. Base flows: Patric and Gould, *Shifting Land Use.*

61. States: David Nye, *Second Creation,* 1. Toqueville: Ibid., 9. Nick: Dec. 5, 1856.

62. Neutral: Chura, *Land Surveyor,* 158. Luria *Geopoetics,* 128 wrote that his "view of the river surpassed the narrow objectives of the lawsuit . . . it showed too much." Grasp: Brian Donahue, "Dammed at Both Ends and Cursed in the Middle: The 'Flowage' of the Concord River Meadows 1798–1862," *Environmental Review* 13 (1989): 65.

63. Lawyers: Oct. 12, 1858.

1. Moccasin Print

1. Henry: He unofficially changed his name after college. Shotgun: Edward Emerson, *Thoreau: As Remembered by a Young Friend* (Boston: Houghton Mifflin, 1912; reprint 1968), 15. Heavy: Michael Sims, *The Adventures of Henry Thoreau* (New York: Bloomsbury, 2015), 61.

2. Relics: May 2, 1859, PT, 67. Revealed: Oct. 15, 1858.

3. Pawed: June 13, 1854. Littered: Oct. 17, 1859, PT, 67. Again: Mar. 28, 1859, PT, 313.

4. James Deetz, *In Small Things Forgotten: An Archaeology of Early American Life* (New York: Doubleday, 1977).

5. These approximate ages are usually expressed as calibrated calendar years before present, to differentiate them from uncalibrated radiocarbon years before 1950.

6. Shirley Blancke and Barbara Robinson, *From Musketaquid Concord: The Native and European Experience* (Concord, MA: Concord Museum, 1985); Shirley Blancke, "The Archaeology of Walden Woods," in *Thoreau's World and Ours: A Natural Legacy,* ed. Edmund A. Schofield and Robert C. Barron (Golden, CO: North American Press, 1993), 242–253. My use of the term "Anthropocene" is unofficial; as noted in the introduction, I use it for materials associated with European contact.

7. Nobler: Mar. 23, 1856.

8. Maker: Jan. 7, 1855. Kneepan: July 31, 1855. Monograph: Apr. 1, 1859. Utility: Nov. 29, 1853. Sea-shore: Mar. 27, 1857. Hachured: Aug. 22, 1860, PT, 88.

9. Shaped them: Mar. 28, 1859, PT, 313. Eternity: Ibid, 314.

10. Stream: Sept. 9, 1853. Creel: June 26, 1852. Feet long: Mar. 20, 1858. Wigwams: Mar. 13, 1859, PT, 269.

11. Ledum: Oct. 23, 1858.

12. Long pieces: Oct. 23, 1855. Blue clay: Aug. 28, 1854.

13. Diluvial: Nov. 14, 1857. Catastrophic: Carl Koteff, "Glacial Lakes near Concord, Massachusetts," *U.S. Geological Survey Professional Paper 475-C (1963),* C142–C144. This lake was given different names as it fell to progressively lower stages, controlled by different outlets. They shared the same bottom sediments.

14. Reaches: Feb. 3, 1855, recalling a memory from Jan. 31, 1855.

15. Shallow lakes: Koteff, "Glacial Lakes." Tectonic: E-an Zen et al., *Bedrock Geologic Map of Massachusetts (Scale 1:250,000)* (Washington, DC: U.S. Geological Survey, 1983).

16. Sediment: July 5, 1852.

17. Geological: Aug. 13, 1858. Irregular line: To the west of the watershed, as Thoreau outlined it, were the "hills of Bolton . . . the height of land between the Concord and Nashua" (June 3, 1850). To the southwest were "the Assabet Hills,—rising directly from the river . . . the highest I know rising thus," beyond which the "convexity of the earth conceals the further hills" (Sept. 24, 1851). Also to the southwest were the "Worcester hills" in Marlborough (Oct. 26, 1853). To the south were the "hills of Hopkinton to Whitehall Pond, the source" of the Concord River (June 11, 1854).

18. Born and dwell: Mar. 28, 1858. Nobscot: May 22, 1853. Note that Thoreau located the Assabet outside of Musketaquid.

19. Mist: May 7, 1852. Grain: Nov. 27, 1857. Southwest: Nov. 16, 1850.

20. Matter: WK, 8.

21. Fan-delta: Janey R. Stone and Byron D. Stone, "Surficial Geologic Map of the Clinton-Concord-Grafton-Medfield 12-Quadrangle Area in East Central Massachusetts, Scale 1:24,000," U.S. Geological Survey Open File Report 2006-1260A [DVD], 2006. Coarse stones: July 20, 1859, PT, 165.

22. Tundra: W. Wyatt Oswald et al., "Post-Glacial Changes in Spatial Patterns of Vegetation across Southern New England," *Journal of Biogeography* 34 (2007): 900–913. Juicy: WK, 16.

23. Nature's economy: Aug. 25, 1856. Summer floods: Aug. 8, 1856.

24. Tilted down: J. Walter Goldthwait, "The Sand Plains of Glacial Lake Sudbury," *Bulletin of the Museum of Comparative Zoology at Harvard College* 42 (1905). See also Carl Koteff et al., "Delayed Postglacial Uplift and Synglacial Sea Levels in Coastal Central New England," *Quaternary Research* 40 (1993): 46–54. They confirmed a tilt of about 0.85 m/km. Continuing uplift during the late Holocene tilts the crust about one millimeter per year. W. R. Peltier, "Global Glacial Isostasy and the Surface of the Ice-Age Earth: The ICE-5G (VM2) Model and GRACE," *Annual Reviews of Earth and Planetary Science* 32 (2004): 111–149 and fig. 21. Acres: "The Dam at North Billerica," *New England Farmer* 14 (1862): 254–255.

25. Rocky spot: May 2, 1859, PT, 67. Midden: Blancke, *Walden Woods*, 242.

26. Fatal illness: Oct. 29, 1837, P1, 9. 1844, after Aug. 1, P2, 103. Aug. 22, 1860. Blancke and Robinson, *Musketaquid*. Remains: Henry Thoreau, "Natural History of Massachusetts," *Dial* 3 (1842): 38.

27. Many more: July 3, 1852. Rocks there: Aug. 1, 1855. Deep mud: Aug. 2, 1859, PT, 215.

28. Indigenous: Dec. 3, 1853. Utensils: Aug. 26, 1854.

29. Warmth: Brian Donahue, *The Great Meadow: Farmers and the Land in Colonial Concord* (New Haven, CT: Yale University Press, 2004), 37–39, provides a good review of environmental archaeology for this interval.

30. Sand is ravished: Feb. 28, 1855. Thick: Aug. 22, 1860, PT, 87.

31. Living in each: Donahue, *Great Meadow*, 41.

32. Disarray: Donahue, *Great Meadow*, 39. Sachem: Blancke and Robinson, *Musketaquid*, 19. Christians: Lemuel Shattuck, *History of the Town of Concord; Middlesex County, Massachusetts, from Its Earliest Settlement to 1832* (Boston; Russell, Odiorne, 1835), 2. A broader background for this transition is given by Charles C. Mann, *1491: New Revelations of the America's before Columbus* (New York: Alfred A. Knopf, 2005).

33. Came from: TP, July 16, 1850. Nobility: Ibid.

2. Colonial Village

1. Sithe: William Wood, *New England's Prospect* (Boston: Prince Society, 1639; reprint, Boston: John Wilson and Son, 1865), 12. See also Ann Zwinger and Edwin Way Teale, *A Conscious Stillness: Two Naturalists on Thoreau's Rivers* (Amherst: University of Massachusetts Press, 1984), 124.

2. Discovery: Lemuel Shattuck, *History of the Town of Concord; Middlesex County, Massachusetts, from Its Earliest Settlement to 1832* (Boston: Russell, Odiorne, 1835), 4.

3. WK, 1.

4. Abused: WK, 8.

5. Mead: Josiah G Abbott, *Argument of Hon. Josiah G. Abbott, on Behalf of the Petitioners before the Joint Special Committee . . . 1862* (Boston: Wright and Potter, 1862), 8. Fen: Sep. 4, 1851. For wetland differentiation, see National Research Council, *Wetlands: Characteristics and Boundaries* (Washington D.C.: National Academy Press, 1995).

6. Inhabitants: Brian Donahue, *The Great Meadow: Farmers and the Land in Colonial Concord* (New Haven, CT: Yale University Press, 2004), 77. Deed: Shirley Blancke and Barbara Robinson, *From Musketaquid to Concord: The Native and European Experience* (Concord, MA: Concord Antiquarian Museum, 1985), 20.

7. Beer: Jan. 24, 1855.

8. Jugular: Nov. 16, 1837, P1, 12. Animal life: Aug. 30, 1853. Times around: WK 6. Globe: Mar. 17, 1859, PT, 264–265.

9. Muck is an official soil type composed predominantly of mineral matter—silt, clay, and very fine sand—mixed with particulate organic matter in high concentrations. Muck contrasts with peat, which consists largely of plant residues that grew in place. Economy: Donahue, *Great Meadow,* xv. Manuring: Ibid., 60. Fodder: Brian Donahue, "Dammed at Both Ends and Cursed in the Middle: The "Flowage" of the Concord River Meadows, 1798–1862," *Environmental Review* 13 (1989): 50.

10. The eight links are river, muck, hay, fodder, manure, fertilizer, grain, and bread. Layout: Donahue, *Great Meadow,* 79. Village: Ibid., 79. Sustainable: Jeremy L. Caradonna, *Sustainability: A History* (New York: Oxford University Press, 2014).

11. Peevish: Donahue, *Great Meadow,* 7. Bridges: Shattuck, *History,* 205.

12. Festival: Jul. 29, 1853.

13. Abate: Shattuck, *History,* 15. Advantage: Ibid. Sewers: JCR, Appendixes, xxviii.

14. Capabilities: Shattuck, *History,* 16. Practicable: *Boston Daily Traveller,* Aug. 17, 1859, n.p., in TP.

15. Unuseful: Shattuck, *History,* 14–15. Denied: Ibid., 15. Inhabitants: Ibid., 16–17. Of the towns: Ibid, 18. See also Abbott, *Argument,* 17.

16. Bridges: Shattuck, *History,* 205.

17. Read: Dec. 8, 1859, PT, 158.

18. Wilder men: May 5, 1859, PT, 76. Your own: Oct. 22, 1857. See Richard Ruland and Malcolm Bradbury, *From Puritanism to Postmodernism: A History of American Literature* (New York: Viking, 1991), 7–15.

19. Come across: Mar. 28, 1859, PT, 311. At last: Oct. 15, 1858.

20. Ashes: Aug. 3, 1852. Creditably: WK, 36. Earth: Sept. 15, 1860, PT, 124.

21. Again: Jan. 27, 1852.

22. Country: Nov. 21, 1851. Pasture: WK, 55. House: Dec. 25, 1856. Complete ruin: Apr. 16, 1860, PT, 135.

23. Landscape: Feb. 5, 1854. Cinnamon: 1849, after Sept. 11, P3, 23. Fires: Nov. 14, 1850.

24. Trout: After July 24, 1846, P2, 265. Indian: Jan. 29, 1856.

25. Therefore: After Apr. 29, 1850, P3, 56. Energy: After Apr. 26, 1850, P3, 71.

26. Already gone: Sept. 8, 1854.

27. Indispensible: JCR, 207.

28. Discharge: The watershed is about 339 square miles above this point, the sum of the Assabet (177 sq. mi.) and the Sudbury (162 sq. mi.) watersheds. Riuer [River]: Henry A. Hazen, *History of Billerica, Massachusetts, with a Genealogical Register* (Boston: A. Williams, 1883), 228.

29. Granted: Hazen, *Billerica,* 228.

30. Defend: Ibid., 278.

31. Construction: Shattuck, *History,* 201. Thereof: SCR, Appendix, v. Said affair: ibid.

32. Shattuck, *History,* 203.

33. *Founding Fish:* John McPhee, *The Founding Fish* (New York: Farrar, Straus and Giroux, 2003). Northwest: David Montgomery, *King of Fish: The Thousand-Year Run of Salmon* (Cambridge, MA: Perseus Books, 2003), 91. Cheape: Ibid., 93.

34. Law: Montgomery, *King of Fish,* 96.

35. Anonymous, "The Dam at North Billerica," *New England Farmer* 14 (1862): 254–255.

36. Haynes: JCR, Appendix H, xxviii.

37. Richardson chain of title: Hazen, *Billerica,* 279.

38. Wayland: JCR, xxxv. Nothing: JCR, xxxvi.

39. Solebat: Theodore Steinberg, *Nature Incorporated: Industrialization and the Waters of New England* (Cambridge: Cambridge University Press, 1991), 141. Reasonable: Ibid., 144–145.

3. American Canal

1. Unintelligible: JCR, 71. Henry F. French, *Argument of Hon. Henry F. French of Boston, March 12, 1862, before the Joint Committee . . . 1860,* CFPL pamphlet C.PAM.60 Item B3, 1862.

2. Privilege of 1704: JCR, xlviii.

3. Boatable: JCR, 70. Causeway: It's near the upstream limit of Musketaquid. Excavations: JCR, 70, quoting Judge French. Strengthened: JCR, 338.

4. Navigation: Lemuel Shattuck, *History of the Town of Concord; Middlesex County, Massachusetts, from Its Earliest Settlement to 1832* (Boston: Russell, Odiorne, 1835), 201.

5. Lower dam: JCR, 68. Cheaply: JCR, 70. Common pleas: JCR, 25

6. Parties: JCR, 181. Defendants: JCR, 25. Ashamed: JCR, 181.

7. Eruption: Gillen D'Arcy Wood, *Tambora: The Year That Changed the World* (Princeton, NJ: Princeton University Press, 2013). Removal: JCR, 199–200.

8. Act: JCR, Appendix K (of the original). Effected: JCR, 26–27.

9. Stream: William Ellery Channing, *Thoreau the Poet Naturalist with Memorial Verses,* new edition, ed. F. B. Sanborn (Boston, Charles Goodspeed, 1902), 3.

10. Concord: Dec. 27, 1855.

11. Baptized: Dec. 27, 1855. Water: Channing, *Poet Naturalist,* 6. Joseph: Michael Sims, *The Adventures of Henry Thoreau: A Young Man's Unlikely Path to Walden Pond* (New York: Bloomsbury, 2015), 16. Outdoor life, 1842–1844: P2, 5. Ecstasy: July 16, 1851. See also Walter Harding, *The Days of Henry Thoreau: A Biography* (New York: Dover, 1982; orig. 1965), 295. Knowledges: WK, 21.

12. Sluggish river: Shattuck, *History,* 196–197. Still level: JCR Appendix, li.

13. Three inches: JCR, 14.

14. Heard: JCR, 175. From 1816: JCR, 175.

15. 1798: JCR, 338. Creation: JCR, 27. Because the water was let out early, the measured fall of "four feet and three inches" between Wayland and Billerica dam was actually greater than normal, thereby unfairly strengthening the case for the dam owners.

16. Bricks: WK, 260.

17. Barge: WK, 263. Feet: WK, 262–263.

18. Landsman: WK, 262.

19. Lived somewhere: WK. Emphasis in original.

20. River project: JCR, 179–187 (detailed descriptions of the 1834 experiments with a method nearly identical to those of 1859).

21. Biography: Refer to Harding, *Days,* Richardson, *Thoreau,* and Walls, *Seeing New Worlds.* Reminiscence: WA, 185.

22. Platform: Manuscript of childhood reminiscence, Morgan Library, cited in Harding, *Days,* 70. Description: Franklin Sanborn, *Henry David Thoreau,* American Men and Women of Letters Series (New York: Chelsea House, 1980), 153–154.

23. Out again: JCR, 97–98.

24. Captain Thoreau: Henry David Thoreau, *The Correspondence of Henry David Thoreau,* ed. Walter Harding and Carl Bode (Westport, CT: Greenwood Press, 1974; orig. 1958), 8.

25. Runs Deepest: Nov. 9, 1837, P1, 10.

26. Neptune: May 3–4, P1, 45 and Harding, *Days,* 59. *Cinderella:* May 13, 1838, P1, 46.

27. Expected: Sept. 16, 1838, P1, 56.

28. Fair Haven: Dec. 15, 1838, P1, 59. Drag: Aug. 13, 1838, P1, 51.

29. Diameter: June 11, 1840, P1, 124–126. Center: 1842–1844, P2, 7. A replica of that boat was built for the Concord Museum and is described in Stan Grayson, "The Musketaquid Mystery: In Search of Thoreau's Boat," *Wooden Boat Magazine* 186 (2005): 39–45.

30. Inspired: July 18, 1839, P1, 77.

31. Middlesex: WK, 70. June 11, 1840, P1, 124–125. Entered the canal: WK, 71.

32. Law: JCR, appendix, ciii

33. Tide: July 6, 1840, P1, 151. Wind: Oct. 6, 1840, P1, 185. Friendship: Oct. 19, 1840, P1, 191. Sail away: Jan. 13, 1841, P1, 220. Hands: Feb. 21, 1841, P1, 271. Crusoe: Ibid. Anchorage: June 2, 1841, P1, 312. Winter: Apr. 27, 1843, P1, 456.

34. Breeze: June 30, 1840, P1, 145. Seaweed: July 6, 1840, P1, 151. Obstacles: Oct. 6, 1840, P1, 185–186.

35. Brink: Dec. 12, 1840, P1, 199. Ridges: Apr. 7, 1841, P1, 297. Any step: Apr. 19, 1843, P1, 454.

36. Our life: Sept. 5, 1841, P1, 331. Fate: WK, 9. Rivers of ore: WK, 414.

37. Nathaniel Hawthorne, *The American Notebooks* (New Haven, CT: Yale University Press, 1932), 165–166, cited in Harding, *Days,* 138. Pacific: WA, 312.

38. Fen: Thoreau, "Natural History of Massachusetts," *Dial* 3 (1842): 19.

39. Passage: Thoreau, "A Winter Walk," *Dial* 4 (1843): 221.

40. Earth: Ibid., 211–226, 221.

41. Once: Thoreau, "Paradise (to Be) Regained," *United States Magazine and Democratic Review* 13 (1843): 452.

4. Transition

1. Emerson: Editor's introduction, P2, in Henry David Thoreau, *Journal 2: 1842–1848,* ed. Robert Sattelmeyer (Princeton, NJ: Princeton University Press, 1984), 445. Damned rascal: Thoreau recounted the fire of Apr. 30, 1844 in his journal on May 31, 1850, P3, 77.

2. Thirties: After Aug. 1, 1844, P2, 104.

3. Hitherward: WK, 35. Clear again: Ibid.

4. With thee: WK, 40. Perfected: WK, 71.

5. Resource: WK, 3.

6. Proprietors: WK, 41.

7. Year round: WK, 40–41. By "English," he was referring to English hay, which was seeded to replace the native vegetation.

8. The phrase "meadowland farmers" refers to those who cut natural hay. Most had mixed holdings, with upland fields, pastures, and orchards as well. Soil: WK, 58.

9. Ants: July 18, 1859, PT, 158. Tool: Mar. 28, 1857.

10. Inch: Aug. 17, 1851.

11. Event: July 16, 1850. Sleep: Aug. 5, 1854. Haymakers: Aug. 7, 1854.

12. Fork it up: Aug. 5, 1854. In the meadow: Sept. 14, 1854. Haying: Aug. 17, 1851.

13. Cleansed: WA, 3. Robert Gross, "'The Most Estimable Place in All the World': A Debate on Progress in Nineteenth-Century Concord," *Studies in the American Renaissance* (1978): 1–15, provides the context for this transformation.

14. Coichuate: JCR, 200. Framingham: See Ann Zwinger and Edwin Way Teale, *A Conscious Stillness: Two Naturalists on Thoreau's Rivers* (Amherst: University of Massachusetts Press, 1984), 70.

15. Good: JCR, 157.

16. Avalanches: JCR, 18. Jaws: JCR, 19.

17. Texas: Harding, *Days*, 177.

18. Chimney: Barksdale W. Maynard, *Walden Pond, a History* (New York: Oxford University Press, 2004), 61.

19. Absurdity: Letter from Ralph Waldo Emerson to William Emerson dated Oct. 4, 1844, cited in Maynard, *Walden Pond*, 60.

20. Concord River: P2, 445.

21. To write: P2, 449.

22. Muck: 1845–1846, Fall Winter, P2, 133. Forest paths: 1846, after Feb. 22, P2, 227. Haven lake: 1845–1846, Fall Winter, P2, 141.

23. Hut on one: WK, 304. Continent: Ibid.

24. Maine Woods: Thoreau, *The Maine Woods: A Fully Annotated Edition,* ed. Jeffrey C. Cramer (New Haven: Yale University Press, 2009).

25. Publish: Harding, *Days,* 243-246.

26. Inventory: Oct. 28, 1853.

27. For modern commentary on this journey and the book, see John McPhee, "Introduction," in Henry David Thoreau, *A Week on the Concord and Merrimack Rivers,* ed. C. F. Hoyde, W. L. Howarth, and E. H. Witherell (Princeton, NJ: Princeton University Press, 2004), ix-xlvi, and David Leff, *Deep Travel: In Thoreau's Wake on the Concord and Merrimack* (Ames: University of Iowa Press, 2009).

28. Purchase: Henry David Thoreau, *The Correspondence of Henry David Thoreau,* ed. Walter Harding and Carl Bode (Westport, CT: Greenwood Press, 1974 [orig. 1958]), 249.

29. Own window: David Wood, personal communication, 2015.

30. Giant: Oct. 27, 1857. Botany: July 23, 1856.

31. Bradley P. Dean, "Introduction," in Thoreau, *Wild Fruits: Thoreau's Rediscovered Last Manuscript,* ed. Bradley P. Dean (New York: Norton, 2000), x-xi.

32. Wanted out: JCR, 44. Pieces: JCR, lv-lvi. River: JCR, lxi. Obligation: Ibid. This request was not completed until October 21, 1859.

33. Increasing it: JCR, 29.

34. Half the shares: Theodore Steinberg, *Nature Incorporated: Industrialization and the Waters of New England* (Cambridge: Cambridge University Press, 1991), 99. Lawrence: Ibid., 103.

5. Port Concord

1. *Beagle:* Charles R. Darwin, *The Voyage of the Beagle* (New York: Harper and Brothers), 1845.

2. Adventure: Henry had previously traveled by steamer and sail to reach Castine, Maine, but he was then a passenger on public transport. Rocky bends: July 26, 1851.

3. Seals on flat: July 29, 1851. Favor you: Ibid.

4. Vessel: July 30, 1851. New Englander: July 31, 1851.

5. Breaks on it: Aug. 26, 1851. Boat for a sail: Oct. 15, 1851. We sail: Jan. 2, 1859.

6. Forests: Aug. 31, 1851.

7. That sea: Sept. 9, 1851.

8. Themselves: Sept. 27, 1851. Our river: Oct. 15, 1851.

9. Reality: May 30, 1853. Large sails: July 21, 1853. Adventure: May 3, 1857.

10. On the sea: Oct. 26, 1851. Middlesex: Nov. 7, 1851.

11. Separate parts: Nov. 9, 1851. Dead ahead: Oct. 15, 1851. Corner Bridge: July 1, 1852. The shore: June 14, 1853. Newspaper: Aug. 24, 1858.

12. The banks: Apr. 15, 1852.

13. Most of the way: Aug. 31, 1852. With me: Ibid. Landscape: Ibid.

14. Complete adventure: Ibid.

15. Boatman: Dec. 2, 1852. Rain: Dec. 27, 1852 ("I took my new boat out"); Dec. 28, 1852 ("Brought my boat from Walden in rain").

16. Repair: Mar. 14, 1853. Dirt: Robert Foley, "Paint in 18th-Century Newport," Newport Restoration Foundation, Newport, RI, 2016, 2.

17. Flat-bottomed: Mar. 19, 1853. Stern: Mar. 22, 1853. Bad sailer: Mar. 23, 1853.

18. His boat: May 7, 1853. Stolen his seat: Apr. 19, 1858. Been stolen: Aug. 22, 1858.

19. Synonyms: May 10, 1854. Possible: Mar. 28, 1854. New one: Oct. 21, 1858.

20. Widest: Mar. 23, 1853.

21. Countries: Mar. 23, 1853.

22. Rapids: Feb. 24, 1855. See Gleason's map at the back of the 1906 *Journal*.

23. Remember: Aug. 6, 1858. Assabet: July 31, 1859.

24. Palm leaf: After Apr. 18, 1846, P2, 237–238.

25. Climes: May 9, 1854.

26. Times: Dec. 5, 1856.

27. Kayack: Feb. 19, 1857.

28. In decreasing order of frequency: William Ellery Channing, Loring, C. and pup, Stedman Buttrick, Sam Pierce, Alcott, Mrs. Wilson and son of Cincinnati, George Bradford, John L., J. Moore, F. Brown, two ladies, Mr. Austin, Sophia, Moncure D. Conway, Hopestill Brown, Tuttle, E. Hoar and Mr. King, Sophia and Cynthia, Smith and Brooks, Daniel Weston, William Tappan, Sophia and Aunt, Rickertson and C., Cabot, Blake, Blake and Brown, Sanborn, Russell, berrying party of ladies, RWE, Frost, Miss Francis and Miss Mary Brown, Father, Willar Farrar, neighbor, Pratt, D. Shattucks, Brown &

Rogers, Loomis and Wilde, Bradford and Hoar, Watsons, Sal Cummings, Mr. Gordon, E. Hoar, Boaz Brown, Thatcher, Mr. Gordon, Elijah Wood, E. Bartlett and E. Emerson, Irishman, some ladies, Stow, Rice.

29. Southwest: June 11, 1860. Valley: Apr. 12, 1859. Itself: May 17, 1854.

30. Valley: July 9, 1859.

31. About: Oct 27, 1857.

32. Ground: Apr. 10, 1852. Through: Apr. 2, 1852.

33. Cold: Apr. 22, 1856. Candlelight: Nov. 9, 1853.

34. Bail it: Aug. 24, 1858.

35. Remarkable bar: Apr. 29, 1860, PT, 154

36. Labored: Apr. 2, 1852. Cooling: July 8, 1852.

37. Today: Sept. 14, 1854. Speed: Jul. 31, 1859. Round: Oct. 15, 1851.

38. Crab: Aug. 12, 1856.

39. Weeds: July 20, 1853. Hauling: Aug. 1, 1855. Wailing: June 20, 1853.

40. My mast: May 28, 1854. The road: Apr. 8, 1856.

41. River: Dec. 29, 1855. Fences: Feb. 19, 1855.

42. *Volaille:* Jan. 14, 1855. Ordinary: Jan. 31, 1855. Irons: Feb. 7, 1855.

43. Blowing: Feb. 15, 1860. Necks: Feb. 3, 1855. Tack: Ibid.

44. Forward: Jan. 31, 1855. Glide though: Jan. 5, 1855. Five feet: Jan. 15, 1855.

6. Wild Waters

1. Cultivation: Jan. 24, 1855.

2. Incomplete: Mar. 23, 1856.

3. Worster: Donald Worster, *Nature's Economy: A History of Ecological Ideas,* 2nd ed. (New York: Cambridge University Press, 1994), 66. Cronon: William Cronon, *Changes in the Land: Indians, Colonists, and the Ecology of New England* (New York: Hill and Wang, 1983), 3.

4. Green: WA, 170. Spring: Aug. 23, 1860, PT, 90. Scum: Kathryn Schultz, "Pond Scum: Henry David Thoreau's Moral Myopia," *New Yorker,* October 19, 2015. Young: WA, 186.

5. History: Leo Marx, "Walden as Transcendental Pastoral Design," in *Walden and Resistance to Civil Government: Authoritative Texts, Thoreau's Journal,*

Reviews and Essays in Criticism, ed. William Rossi, 2nd ed. (New York: W. W. Norton, 1992), 377–390. Cronon, *Changes.* Engineering: Barksdale W. Maynard, *Walden Pond, a History* (New York: Oxford University Press, 2004), 255–263.

6. Based on Thoreau's measured discharge of the Concord River versus the flow-through rate of Walden Pond.

7. Ruralist: Buell, *Imagination,* 39. Pursuits: Ibid., 25. Outskirts: Laura Dassow Walls, *Seeing New Worlds: Henry David Thoreau and Nineteenth-Century Natural Science* (Madison: University of Wisconsin Press, 1995), 266. Richard Judd, *Second Nature: An Environmental History of New England* (Amherst: University of Massachusetts Press, 2014).

8. Vegetable: Nov. 4, 1852.

9. Shoveling manure: Oct. 4, 1857. Landless: Robert Gross, "'That Terrible Thoreau': Concord and Its Hermit," in *A Historical Guide to Henry David Thoreau,* ed. W. E. Cain (New York: Oxford University Press, 2000), 193. Advanced guard: Daegan Miller, "Witness Tree: Landscape and Dissent in the Nineteenth-Century United States," Ph.D. dissertation, Cornell University, 2013. Wand: Andro Linklater, *Measuring America: How the United States Was Shaped by the Greatest Land Sale in History* (New York: Penguin, 2003), 5.

10. Frequency of Thoreau's surveying is based on tallies from the online archive of the Concord Free Public Library. Properties: CFPL Special Collections, Thoreau's Surveys 25a, 42, 77, and 84. On New Year's Day in 1851, Thoreau's mechanical interest brought him to Nashua River west of Concord. Fall: Sourced in Robert D. Richardson Jr., *Henry Thoreau: A Life of the Mind* (Berkeley: University of California Press, 1986), 228.

11. Broadside: Reproduced in Patrick Chura, *Thoreau the Land Surveyor* (Gainesville: University Press of Florida, 2010), 85 from the Berg Collection, New York Public Library. Satan: Henry David Thoreau, "Walking," in *Excursions* (Boston: Ticknor and Fields, 1863), 161–214, 170.

12. Worth the while: July 24, 1852.

13. Darkness: Nov. 11, 1850. Thoreau, "Walking." Contrition: Patrick Chura, *Thoreau the Land Surveyor.* Miller, "Witness Tree."

14. Wildness: Aug. 30, 1856.

15. Being quite moody, Thoreau occasionally complained about his emasculated and demoralized world, but this was usually in the context of religion or humor.

16. Vain to dream: Aug. 30, 1856.

17. Landscapes: Leo Marx, *The Machine in the Garden: Technology and the Pastoral Ideal in America* (New York: Oxford University Press, 1964), 390, mirrored that thought: "The location of meaning and value does not reside in the natural facts or in social institutions or in anything 'out there,' but in consciousness," in the "mythopoetic power of the human mind. Cronon, *Changes,* does not cite Thoreau's precedent. Within an hour: Jan. 26, 1853.

18. Genius: Feb. 16, 1859. W. E. Cain, "Henry David Thoreau, 1817–1862: A Brief Biography," in *A Historical Guide to Henry David Thoreau,* ed. W. E. Cain (New York: Oxford University Press, 2000), 4. Source of Whitman quote is Horace Traubel, *With Walt Whitman in Camden, 1905–1906* (New York: Rowman & Littlefield, 1961), 3:375. Tameness: Nov. 16, 1850.

19. World: Thoreau, "Walking," 183.

20. Lakes: After Mar. 11, 1845, P2, 124. Every direction: Ibid., 126.

21. Dry land: Apr. 21, 1852. Otherwise: Apr. 23, 1852.

22. Sails: Apr. 15, 1852.

23. Landscape: Mar. 8, 1853.

24. Grass: Mar. 16, 1859.

25. Without end: May 30, 1853. Northward: Mar. 16, 1859. Has come: Ibid. Any month: Mar. 19, 1859, PT, 275.

26. Aslant: Mar. 19, 1859, PT, 276. Your boat: Mar. 28, 1859, PT, 323.

27. North: Apr. 19, 1852. Ponds: Apr. 22, 1857.

28. The deep: May 8, 1854. See also May 6, 1854.

29. Rise: Aug. 5, 1854. Safely: July 16, 1850. Down-stream: Aug. 22, 1856.

30. Floated: Oct. 27, 1857. Hands: Mar. 19, 1859, PT, 275.

31. Inspiriting: Oct. 27, 1857. Inches: Nov. 11, 1853. One side: Nov. 14, 1853.

32. Undulation: Dec. 4, 1856.

33. Blue water: Mar. 12, 1854. Field: Apr. 16, 1852. Would be: Apr. 16, 1852. Voyages: Apr. 15, 1855.

34. Reflections: Apr. 15, 1855. Heavens: Apr. 9, 1856. Fallen sky: June 11, 1851. Falling: Mar. 2, 1860. Concealed: Nov. 6, 1853.

35. Waters: Mar. 16, 1859. Faces: May 7, 1854.

36. Landscape: May 6, 1854. Earth: Mar. 16, 1859. Mapped: Mar. 17, 1859.

37. Shore: May 7, 1854. Pleasure: Apr. 24, 1852. Hills: May 7, 1854. Tingeing: Nov. 20, 1853. Oct. 31, 1853. Curving line: Mar. 23, 1859, PT, 293. Surprising: Mar. 17, 1859, PT, 269. Fjords: Mar. 16, 1859.

38. Element: Nov. 7, 1853.

39. Gray: Nov. 14, 1853. Light-blue: May 7, 1854. Nature: Oct. 26, 1857.

40. Cut-grass: Aug. 1, 1856.

41. Meadow: July 14, 1855. Gale: May 7, 1854. Sea-beach: Apr. 14, 1856. Larger ones: May 7, 1854. Over them: July 11, 1856. Prows: Nov. 14, 1853.

42. Long time: Mar. 16, 1859. Smell: May 4, 1859, PT, 69. Beaches: May 19, 1854. Scene: Aug. 29, 1854. Streams: July 22, 1859, PT, 175.

43. Shore: Aug. 3, 1859, PT, 221. Lost in it: Aug. 24, 1858. Hastening: July 4, 1860, PT, 293.

44. Waves run: July 31, 1859, PT, 203.

45. Bell: Dec. 29, 1854. Glass: June 27, 1856. Decks: Jan. 26, 1857.

7. River Sojourns

1. Profession: Sept. 7, 1851.

2. Whitish: Oct. 16, 1859, PT, 60. Ready set: Aug. 22, 1858. Overboard: Aug. 22, 1858. Bull: Aug. 22, 1858.

3. Bride: Apr. 23, 1857. Unchanged: Mar. 28, 1858. Want: Nov. 1, 1858.

4. Communicable: Aug. 29, 1858.

5. Advantages: Oct. 17, 1859, PT, 67. Transplant: Nov. 20, 1857. Native soil: Nov. 20, 1857. Africa: Mar. 18, 1858.

6. Valley: Sep. 13, 1856.

7. The methodology section for the statistical analysis of Thoreau's journaling frequency is posted online at http://robertthorson.clas.uconn.edu/writing/books/the-boatman/online-repository/.

8. Statistics for modern boating are from Statista, 2016, www.statista.com/statistics/240522/recreational-boating--average-annual-boating-days-in-the-us, accessed July 27, 2016. The values are: kayaks / canoes, 11.2 days per year; sailboats, 10.7; and general personal watercraft, 10.7.

9. Transparent: WA, 300. Fetters: Nov 25, 1850. Piers: Jan. 22, 1855. Apart: Feb. 28, 1855. Ice: Feb. 17, 1855. Scene: Feb. 28, 1855.

10. Lying flat: Apr. 10, 1856.

11. Blotches: Mar. 2, 1860, PT, 42. Heights: May 1, 1859, PT, 64.

12. Fishes: May 28, 1854.

13. Quire: May 12, 1854.

14. Fireflies: June 15, 1852.

15. Ephemerae: June 2, 1854.

16. Cat briars: Jan. 12, 1854. Roars: May 30, 1857.

17. Dragonflies: June 13, 1854. Hot: July 3, 1854.

18. Furnace: July 9, 1852.

19. Water walks: July 10, 1852. Emperor: July 10, 1852.

20. Snap: Aug. 26, 1854.

21. Water lily: Jun. 16, 1854.

22. Detect me: Aug. 24, 1858.

23. When I was completing this manuscript in August 2016, Concord's three rivers were in a state of extreme drought, with the Assabet at its lowest level in thirty-six years. Mark Pratt, "Small Part of Massachusetts Said to Be in 'Extreme Drought,'" *Chicago Tribune*, August 11, 2016.

24. Dress: Aug. 14, 1859, PT, 237.

25. Blow: Aug. 31, 1852. Waters: Sept. 27, 1858. Adhesive: Sept. 4, 1854. Lime: July 30, 1856.

26. Heaven: WA, 274.

27. Flag: Sept. 27, 1857. Boats: Nov. 1, 1852. Skiffs: Oct. 17, 1856.

28. Elysian: Oct. 6, 1851.

29. Sail: Aug. 19, 1853.

30. Sail: Nov. 6, 1855. Stove: Feb. 6, 1855 ("the stove all day in my chamber").

31. Another: Nov. 9, 1853. Hope: Oct. 13, 1859, PT, 44.

32. Summer: Nov. 11, 1858. Ice: Nov. 11, 1858. Hands: Nov. 9, 1853. Row: Dec. 2, 1852. Icicles: William Ellery Channing, *Thoreau the Poet Naturalist with Memorial Verses*, ed. F. B. Sanborn, new ed. (Boston: Charles Goodspeed, 1902), 11–12. Inactivity: Dec. 2, 1852.

33. Sharp: Dec. 9, 1856.

34. Otter: Jan. 11, 1855. Snow: Dec. 30, 1840, P1 200.

35. Desert: Feb. 2, 1860, PT, 282.

36. Thick-skinned: WA, 291.

37. Life: Jan. 20, 1856.

8. Consultant

1. Tons: Oct. 4, 1856.

2. Bushels: JCR, 36. Home: Aug. 26, 1858. Lincoln: Ibid.

3. Lost: Aug. 9, 1853. Standing: Aug. 19, 1853.

4. Meeting: Boston Journal, "The Meadow Lands of the Concord River Valley; Meeting of the Proprietors at Concord," *New England Farmer* 11 (1859): 76.

5. Legislature: JCR, 36.

6. Husbandman: JCR, 14.

7. Mellen: JCR, 208.

8. Inaccessible: JCR, 15.

9. Generations: JCR, 20.

10. Subtle fingers: JCR, 20.

11. Pure water: JCR, 31.

12. Perceived: March 7, 1859, PT, 227. Nature: Mar. 8, 1859, PT, 229.

13. Observed fact: Ralph W. Emerson, "Thoreau," in Thoreau, *Excursions,* ed. Leo Marx (Boston: Ticknor and Fields, 1863), 7–33. See also Edward Emerson, *Thoreau: As Remembered by a Young Friend* (Boston: Houghton Mifflin, 1917; reprint, Concord, MA: Thoreau Foundation, 1968), 148.

14. Engrossed: JCR, 449. See also Massachusetts General Court (Legislature), *Acts and Resolves Passed by the General Court of Massachusetts for the Year 1859, First Session* (Boston: William White, Printer to the State, 1860).

15. Heaped up: Mar. 18, 1859, PT, 269. Fall of 1853: Nov. 19, 1853.

16. Mathematically: The variables are width of the abutments, the number and width of each pier, the number of bridges, the mileage between bridge, and the general stage of the river. Tufts: Feb. 17, 1855.

17. Willow rows: Apr. 12, 1859, PT, 28. Steeper: Ibid. Brown: Apr. 1, 1859, PT, 330.

18. Machinations: Apr. 8, 1859, PT, 10.

19. *Assistant:* Walter Harding, *The Days of Henry Thoreau: A Biography* (New York: Dover, 1982, orig. 1965), 409. Revenue: Apr. 3, 1859, PT, 338.

20. Representatives: The Town of Concord was represented by Hon. Simon Brown, Messrs. J. Reynolds, Elijah Wood Jr., and Samuel H. Rhoades; Wayland by Col. David Heard and Mr. Abel Gleason; Sudbury by Messrs. John Eaton and T. P. Fairbanks; Bedford by Mr. J. W. Simonds; Carlisle by Mr. Artemas Skilton; and Weston by Mr. Nathan Baker. For the Talbot Factory, Messrs. C. P. and T. Talbot were represented by Hon. J. G. Abbott and Hon. B. F. Butler, of Lowell, and G. A. Somerby, Esq., of Waltham. For the Faulkner Mills, Messrs. James R. and Charles Faulkner were represented by G. H. Preston, Esq., of Boston. Three committee members present were Parker, Wrightington, and Russell.

21. Sheet of water: JCR, 61–62. Spongy: JCR, 62–63. Flowed: JCR, 63.

22. Questionnaire: The questionnaire in the Thoreau Papers at Concord Free Public Library asks for "Meadow Owners" (name), "No. of Acres," "Meadow, where situated," "Value per Acre before damage." "Present value per Acre," and "Remarks." Also on the committee were John Eaton, of Sudbury; N. O. Reed, of Bedford; Elijah Wood Jr. and Joseph Reynolds, of Concord; Thos. Page, of Carlisle; Abel Gleason, of Stow; Nathan Barker, of Weston; and Thos. J. Damon and R. S. Fuller, of Wayland. Private letters: TP. The letter from Richard Heard, dated Concord, July 8: says, "By the vote of the Committee I am requested . . ."

23. Case: Kenneth W. Cameron, "Thoreau in the Court of Common Pleas," *Emerson Society Quarterly* 14 (1959): 86, cited in Harding, *Days,* 326 n. 5. "Before the case came up for trial, he hired Thoreau to check the height of the dam. Thoreau was promptly served with a summons as an expert."

24. The text is (*x* signifies something illegible): "Baldwin's map 2d one Bacon's survey Richard *xxxx*plan of two things. In Wayland David Heard & other Heard/David S Child/Abel Gle. . . . /In Bedford John W. Simonds/Nathan O. *X*ead/In Billerica Jonathan Hill/In Carlisle S Skelton."

25. Day more: TP, survey notebook.

26. Entries: TP, loose notebook paper.

27. Claims: Patrick Chura, *Thoreau the Land Surveyor* (Gainesville: University Press of Florida, 2010), 15. It's possible that these tasks were specified in the missing pages of the appointment letter. River survey: Sara Luria, "Thoreau's Geopoetics," in *GeoHumanities: Art, History, Text at the Edge of Place,* ed. M. Dear, J. Ketchum, S. Luria, and D. Richardson (London: Routledge, 2011), 128.

28. Prior to: Grove K. Gilbert, *Report on the Geology of the Henry Mountains,* United States Geological and Geographical Survey, Rocky Mountain Region

(Washington, DC: General Printing Office, 1877); Grove K. Gilbert, *The Transportation of Débris by Running Water,* United States Geological Survey Professional Paper no. 86 (Washington, DC: Government Printing Office, 1914).

29. In both dated transects, he differentiated the "westerly, first" (July 7, 1859) and the "dry & hard bank—by the river" (June 16, 1859) from the "wet, or main, part" (July 7, 1859) of the meadow, and from the transition to the flanking sandy soils. Sedges: Brian Donahue, *The Great Meadow: Farmers and the Land in Colonial Concord* (New Haven, CT: Yale University Press, 2004), 92.

30. Shows: Sept. 12, 1851. Truss: Aug. 23, 1852.

31. Know this: Aug. 25, 1856.

32. Drainage: TP. His official datum was the "corner stone on the E side of Hoar's steps," which he measured as "5 feet 6½ inches" above "summer level" in that place at that time. His unofficial, more convenient datum was the "notch in willow at my boat place" (June 27, 1857). In TP is a note linking his local datum at Hoar's steps to a "real world" datum surveyed by railroad engineers: the "E end horizontal part of the iron trusses of the Stone Bridge," which was "3 feet 6⅝ inches higher." Summer level was thus 9 feet, and ¹¹⁄₁₂ of an inch below the bridge truss. Anecdotally, this was just "above Flint's low rock." On June 20 he also tied his datum to a stone located one foot east of Hoar's extreme northwest corner, where his land bounded Whiting's. At the time, the water there was 12½ inches below that stone. Sloshing: Mar. 13, 1857.

33. Overturned: June 18, 1859, PT, 104. The post-flood drop in water level at the Fordway (2 feet 6 inches) was only slightly less than the drop in Concord (2 feet 9½ inches), proving that the former controlled the latter. Thoreau did not cite the numerical data from Saddle Rock, but his journal indicates he kept track of the water transfers.

34. Bridge: Nov. 19, 1853. Fall: Feb. 17, 1855.

35. Grown much: June 22, 1859, PT, 109. Shallow places: June 24, 1859, PT, 111. Feeling: William Ellery Channing, *Thoreau the Poet Naturalist with Memorial Verses,* new ed., ed. F. B. Sanborn (Boston: Charles Goodspeed, 1902), 264. See William L. Howarth, "Concord River Survey (F13)," in *The Literary Manuscripts of Henry David Thoreau,* ed. W. L. Howarth (Columbus: Ohio State University Press, 1974), 303. He identifies the "statistics of bridges."

36. Controversy: June 24, 1859, PT, 114. Begun to fall: June 24, 1859, PT, 114.

37. Jon Hill and David Heard letters: Albert F. McLean Jr., "Addenda to the Thoreau Correspondence," *Bulletin of the New York Public Library* 71 (April 1967): 265–267.

38. Bridges: CFPL, Series II, No. 107c, Box 1, Folder 5, Vault A35, Thoreau no. 11. In the related folders are thirty-five notebook pages, one for each bridge. In the first column is the name of the bridge and any synonyms. The next four columns involve bridge history: "When first created," "When discontinued," "When and how altered, within this century, so as to affect the passage of water," and "Authority for the preceding statement." Four involve the physical architecture: "Material at present," "Width between abutments in feet," "No of Piers," and "Width of each." The final three columns contain the depth at the right, middle, and left abutments, respectively. The significance of one partially illegible column is unknown. One involves transient obstructions, mainly drifted meadow. Saddle Rock: It shows the water rising from 2%12 feet deep at Heard's Bridge in Wayland to 7⅙ feet on the Corner Bridge in Billerica, which is the lowest above the outlet at the Fordway. The same statistic for the 1833 survey is 2.865 feet, with the difference due to the increased height of the dam in 1828. Lakelike condition: July 22, 1859, PT, 177. Thoreau compared between-bridge reaches between 1833 and 1811, indicated by data on a torn sheet of notebook paper approximately 4 by 3 inches in TP: "Names of Bridges on Baldwin's Map of 1811. 14 Bridges."

39. A high-resolution scan of this table is posted as part of an online exhibit at the CFPL titled "Thoreau's Surveys," item 107b.

9. Mapmaker

1. Brook: July 4, 1859, PT, 120. Inches wide: July 7, 1859, PT, 130.

2. The original remains in the Special Collections archive of the Concord Free Public Library as item 107a. High-resolution digital copies are available online from the listing "Thoreau's Surveys" within its department of Special Collections. My versions are simplified and printed on four successive pages (Figures 20A–20D).

3. Channels: Theodore Steinberg, *Nature Incorporated: Industrialization and the Waters of New England* (Cambridge: Cambridge University Press, 1991), 88.

4. The petitioners attached Baldwin's original 1834 map as Plate I in the first appendix.

5. These measurements were given in rods and links, indicating that Thoreau retained the original units. A rod, also known as a perch or a pole, is 16½ feet long or ¹⁄₃₂₀ of a mile. A link is 0.66 feet or 7.92 inches long, because there were 100 links in the length of a Gunter's chain, which was 66 feet long.

6. Most downstream is the one at the base of the Holt. Above that are the lower part of Barrett's Bar, an unnamed bar above the old North Bridge, the Assabet Bar, and at the bend below Clamshell. Thoreau's map shows the height of the

dam as "6½ feet" and the height of the flash board above it as 15.3 inches or 1.275 feet.

7. During this twenty-two-year span the gradient near the mouth of the Assabet between Barrett's (Hunt's) Bridge to the north and the Turnpike Bridge to the south remained negligible, being 0.000 in 1811 and 0.023 in 1834. These before-and-after comparisons of slope changes between bridges cannot be extended to the later 1861 study, because the latter used different stations.

8. Thoreau must have noticed the changes in slope he was recording: from 0.622 to 4.776 feet per mile northerly and from 1.589 to 0.986 feet per mile southerly. Nor could he have missed the fact that the boat-place bar was newly appeared, and he knew this area had recently experienced a surge of sediment accumulation. He does not, however, explicitly link these three observations in his writing.

9. Doodle-like: Sara Luria, "Thoreau's Geopoetics," in *GeoHumanities: Art, History, Text at the Edge of Place,* ed. M. Dear, J. Ketchum, S. Luria, and D. Richardson (London: Routledge, 2011), 133.

10. Width: On his map is the explanation "Breadth on right bank in rods, as paced." I count 292+/- depth measurements. Specifically, they are "soundings, in feet, so much below summer level—which is 5 ft 6½ inches below the wall at Hoars steps—or 2 ft 8 inch below the notch in willow at my boat." In this sentence are three of the eight temporary data stations that Thoreau linked back to a common benchmark for his soundings. Thoreau specifically identifies "hard" and "soft" river bottoms based on his long pole; the names of hills; places as "weeds," "weedy," "very weedy," "sandy," or "rocky"; and "swift" or "deep." In a smaller size, and written in all orientations, are place names that cross-reference the journal. In total, there are four stages to the overlay lettering on the map, not counting the two colors of ink. The Mar. 17, 1859, entry indicates that the actual river was far more complex than his lines indicate. He located "rudely the soft meadow by pencil dots on the map." July 7, 1859, PT, 179.

11. River length: July 31, 1859, PT, 210. The length indicated by the scroll map was about "24 miles 269 rods" from the "broad clear fall" just below Pelham Pond to the Billerica dam. Musketaquid length: July 7, 1859, PT, 130. From Bridle Point Bridge to the dam was "16 miles 48 rods" as the crow flies, and a distance of "22 miles 289 rods" along the path of the river.

12. Reynolds: July 8, 1859. The journal entry for July 31, 1859 (PT, 210) provides a sample of soundings in text: "The average depth between upper end Sudbury canal & B. P. B is 8½+ ft Bet. Sud canal & Heards B 8⅓ Bet. Sud canal & Pelham P. say 9 ft. The last is say 3¹⁄₁₆ miles if then it was on an av. about 8 ft between Sud canal & Falls (as stated the 22d) it is about 8⅛ bet. Pelham Pond & Falls."

13. Brink: May 16, 1854, and July 9, 1859, PT, 133. Lift-off: July 9, 1959, PT, 133. Individual uplifted masses could be up to "200 feet long and 100 feet wide, and often a full 3 feet thick of solid earth." Brink: July 9, 1859, PT, 133; "This might be called Brink bush—or Drift bush River fence." Smooth lawns: Lift-off requires that winter flooding be deep enough to create thick ice. Next is a freeze cold enough to reach down into the turf. Spring flooding must then be deep enough to buoy up the ice and tear up the attached turf. The final step is to float the tufts away like vegetated floes. Concord's land use history is consistent with enhancement of this meta-process: higher and faster spring floods resulted from greater runoff; deeper freezes from close mowing; and a softer bond between turf and muck caused by a rise in water table due to flowage.

14. Billerica: Aug. 25, 1856. This surmise was just about the time he began to study seed dispersion in the pine-oak forests of the upland. Flows: Thoreau's interpretation is supported by hydrologic data on Concord River discharge collected since 1937 by the U.S. Geological Survey. Muster: Aug. 9, 1859, PT, 229. Changed: Aug. 8, 1856: "that our river meadows were historically drier than now originally, or when the town was settled."

15. Meadow: Aug. 26, 1858. Pipe stem: Sept. 30, 1856. See also Mar. 8, 1857.

16. Respectively: The Assabet was 107.25 feet wide and 3.75 feet deep, yielding a ratio of 28.6 to 1, characteristic of broad, sandy streams. Feet: July 16, 1859, PT, 153. Deep: July 16, 1859, PT, 153. Years: July 15, 1859, PT, 135. These descriptions and measurements were based on maps as accurate as the scroll map, and which are also archived in the CFPL.

17. Occasioned: July 20, 1859, PT, 162. Maiden aunts: Robert G. Gross, "Faith in the Boardinghouse," *Thoreau Society Newsletter* 250 (Winter 2005): 1–5. False spring: Cary J. Mock et al., "The Winter of 1827–1828 over Eastern North America," *Climate Change* 83 (2007): 87–115.

18. Quotes are from July 16–18 on PT, 153–155. Thoreau's information was drawn from four draft maps of the Assabet channel considered an addendum to the scroll map. One bears the title "Assabet River July 18 59 Outline of S side of AB [illegible] from Hubbards Plan scale of 20 rd per inch Measured width in two places. C very rudely by pacing." The second is a sketch map of Abel Hosmer and Colburn Farm showing its fathoming depths. Also in the file is a note referring to the scour pool below the one-arch bridge: "Bridge 25 feet wide [pool] 22½." All other depths on the sheet are from 4¾ to 2¼. A good part of the file concerns this bridge and the adjacent lands as sketch mapped. Spring: July 15, 1859, PT, 154. Sandbar: Herbert Gleason's photo 1920.16 in the CFPL shows a still-active sandbar.

19. Different: July 15, 1859, PT, 155. Sawdust: Sept. 5, 1854. Stream: July 15, 1859, PT, 155.

20. Hemlocks: Apr. 1, 1852. River: Mar. 24, 1855. Hill: Aug. 1, 1859.

21. Country: Apr. 1, 1854.

22. Level: Nov. 9, 1855. Pot: Oct. 20, 1856.

23. Coal: Aug. 31, 1856. Hills: Oct. 10, 1857. These interpretations are mine. Thoreau wasn't explicit in his interpretations.

24. Position: George B. Bartlett, ed., *The Concord Guide Book,* ill. L. B. Pumphrey and Robert Lewis (Boston: D. Lothrop, 1880).

25. Gravel: July 7, 1859, PT 155. Junctions: Ibid. Island: Mar. 17, 1859, PT 53.

26. Stuff: Nov. 11, 1855. Lorings: May 1, 1852. Flowed down: Ibid. Rices: June 24, 1859, PT 113.

27. Current: Aug. 22, 1856. Sawdust: Nov. 7, 1855. Lees Bridge: July 7, 1859, PT, 132.

28. Thoreau noted the steepened slope of the lower Assabet channel: July 7, 1859, PT, 132. One Pond: TP.

29. Volume: The mass can be estimated as a the area under the curve of Avery's bathymetric profiles for profile of the 1861 ECR: W. Alvord Daniel, Charles S. Storrow, and Herbert J. Shedd, *Report of Experiments and Observations on the Concord and Sudbury Rivers in the Year 1861* (Boston: William White, Printer to the State, 1862), Plate 2. This estimate assumes an arbitrary base and constant width. Unnamed bar: The name "boat-place bar" is mine because it marks Thoreau's boat place.

30. Addenda written after July 22 are included here to avoid redundancy.

31. Hitchcock: Edward Hitchcock, *Final Report on the Geology of Massachusetts in Four Parts* (Northampton, MA: J. Butler, 1841). Zen E-an et al., *Bedrock Geologic Map of Massachusetts,* U.S. Geological Survey, Scale 1:250,000, 1983.

32. For brevity, I offer this summary of Henry's six reaches: Southernmost is the Pelham Pond Reach, which marks the opening of the alluvial valley. A lake-like reach of wild rank meadows so special they received a special name, the Beaver Hole Meadows. Entering the Sudbury Meadows Reach, the river flattens, widens, and assumes a well-defined meandering pattern. There he found small bays at each bend that were not present on Baldwin's 1834 map, and which were highlighted by stands of bulrush. This suggested that the meanders were no longer active but, in fact, drowned, indicating historic submergence. Below that is the Hill-Lake Reach, centered on Fair Haven Bay. This is the oddest reach in the valley. Though the current is virtually stagnant, there are bedrock cliffs—Conantum, Fair Haven, Lees—standing above narrows so deep that the weeds cannot span them. Leaving Fairhaven Pond, Thoreau

writes of the river "breaking through highlands, issuing from some narrow pass. It imparts a sense of power" (Apr. 11, 1852). Next is the Concord Reach or Rapids Reach, which is aligned along the local bedrock grain and largely filled with sandy gravel from the Assabet River. Downstream from the Holt is the Carlisle or Deep Lake Reach. It's flat, wide, and much straighter than anything to the south, wide enough to require twenty wooden piers on the Carlisle Bridge. After that is the Billerica or Shallows Reach. This is the gradual transition from a deep/muddy channel to a shallow/bedrock channel as the river rises out of its alluvial basin toward the natural bedrock outlet at the Fordway. The final reach is the Fordway and Falls Reach. There the river downcuts the lip of the alluvial basin to create a bedrock rapid, called the Falls, before entering the reservoir of the Billerica dam.

Henry's six-part division of the river follows the bedrock and glacial geology almost exactly. The Pelham Pond reach and the main part of the Sudbury Meadows reach traverse the older, generally higher Proterozoic rocks of the Avalon terrane. The former flows generally parallel to the regional tectonic grain, whereas the latter outlines a series of blocks bounded by brittle faults trending north-northwest. South of Nut Meadow Brook the Hill-Lake Reach flows above a large fault trending north-northwesterly in a valley deepened by glacial flow in that direction. The Rapids reach flows parallel to bedrock strike and the retreating ice margin in a less resistant rock unit dominated by diorite. This anomaly suggests that it was diverted there as a meltwater channel linking the pair of adjacent lakelike reaches. The Carlisle reach is the longest and straightest because it flows through otherwise hard rocks in a segment aligned by a brittle fault. The most northerly Billerica or Shallow reach is underlain by the resistant Andover Granite standing above weaker rocks to the north. This includes the shallows leading up to the Fordway and the falls below it. The basin of Fairhaven is a glacial kettle because it's flanked by collapsed sand and gravel. The present water is impounded by the earlier, deeper portions of the Assabet Delta.

33. Thoreau recognized the presence of the mound from his soundings. It's even more clear from the higher-resolution bathymetry drawn by surveyor John Avery for the 1861 Engineering Commission investigation, part of which is reproduced in this book (Figure 23).

34. Snake: July 5, 1859, PT, 124. Valley: July 22, 1859, PT, 182.

35. Mile: July 5, 1859, PT, 132. Still: July 30, 1859, PT, 191. Fishermen: Aug. 3, 1859, PT, 219.

36. North: July 5, 1859, PT, 123. Current: July 22, 1859, PT, 181. Strained: Aug. 5, 1854.

37. Carriage: July 22, 1859, PT, 181, and June 24, 1859, PT, 114. Concord: Aug. 2, 1859, PT, 214. Bend: July 22, 1859, PT, 181. Canal: Ibid. Stagnant: Ibid.

38. Pedestal: July 22, 1859, PT, 170. "Just above the Fordway" are large boulders that are "curiously water worn" into the shape of a pedestal. "Rocks half a dozen feet in diameter which were originally of the usual lumpish form are worn thus by the friction of the pebbles &c washed against them—by the stream at high water." Ink: Ibid., 171.

39. Lake upstream: And the channel remains inefficient, even during higher floods, because the deeper water raises the velocity, which raises the turbulence, which keeps the velocity in check. Hydraulic dam: "I judge that in a freshet the water rises higher as you go down the river" (July 8, 1859, PT, 134).

40. Rubble: July 22, 1859, PT, 172. Crockery: Ibid., 170.

41. Rapidly: Ibid., 181. Sand: July 7, 1859, PT, 128.

10. Genius

1. Fordway: Specifically, "nearly on a level (say within ¼ of a foot) with the bottom 20 miles above," July 22, 1859, PT 184–185: "If you should lower the river 6 feet uniformly—it would be dry for about 1 mile in Billerica (between middle [133] & Corner Bridges) leaving numerous small lakes & 3 long narrow ones above in Wayland, Concord, & Carlisle." He observed that the basin's notched edge at the Fordway was "not quite 7¾ feet below the level of the meadow surface at the Sudbury Causeway" (July 22, 1859, PT, 186), which is located twenty-two miles distant on the other side of the ancient lake basin.

2. Two diameters: WA, 281. Gains: July 22, 1859, PT, 185.

3. Fall: Jul. 22, 1859, PT, 186. Been: Ibid. At this point he tests his hypothesis. Absolutely: June 22, 1859, PT, 185.

4. Wave: John M. Barry, *Rising Tide: The Great Mississippi Flood of 1927 and How It Changed America* (New York: Simon and Schuster, 1977), 39.

5. Meadows: Apr. 7, 1853. Standstill: July 30, 59, PT, 197. Drowned: Ibid. Stream: Aug. 3, 1859, PT, 219.

6. Created thing: Mar, 7, 59, PT, 227. Ages: Aug. 3, 1859, PT, 217.

7. Bridge: Jul. 24, 1859, PT, 114.

8. Strawberry: July 30, 1859, PT, 198.

9. Curved channel: Jul. 30, 1859, PT, 198.

10. Current: Aug. 2, 1859, PT, 214. Places: Aug. 2, 1859, PT, 214.

11. Minute: July 25, 1859, PT, 189. Rating curve: June 23, 1860, PT, 371. Henry's approximation of a rating curve for channel area and depth is this: "For a river that can rise and fall 'eight and one twelfth feet' below the railroad truss, the

depth in the channel is half again as much as that, and the width varying up to eight or nine times the in-channel width.

12. Commissioners: TP.

13. Methodology: TP. For normal annual rainfall of 44.42 inches per year, the discharge from the Hopkinton Compensating Reservoir (Sudbury River) averaged 622 cubic feet per minute for the period June 13-November 14, 1852. This compares with Thoreau's measurement of 266 cubic feet per minute for the summer flow of the Sudbury. See online repository.

14. Sluggish: July 22, 1859, PT, 174. Former (fastest): June 24, 1859, PT 115. Fall: June 24, 1859, PT, 114. The "swiftest" after the "fall" indicates diffusive decay that proves hydraulic ponding

15. Leveling: JCR, 214.

16. South Bridge, July 27, 1859, PT, 191. He references the earlier passage on July 12. Quickly: JCR, 214.

17. Notch: Aug. 5, 1859, PT, 223: "I have made these observations on the stage of the water within a week—referring it to the top of a stake (which I will call X) ¾ inch above summer level." Dollars: Aug. 14, 1859, PT, 245. His ultra-precision, when compared to the horizontal distance of twenty-two miles, is equivalent to one part in 90 million. This reflects his lifelong habit of making measurements as accurately as humanly possible.

18. Half: Aug. 5, 1859, PT, 225 (includes an addendum from the seventh). Concord: Aug. 5, 1859, PT, 225.

19. Extend datum: Aug. 11, 1859, PT, 245. For this time period, he called his stake datum at X the "hub," which was translated in the 1906 journal as "mark." The result was six high-precision data points (within a range of only 0.3 feet) spanning the three days of August 12, 13, and 14. The limited drop at South Bridge meant that Barrett's Bar was preventing a greater drop, indicating that it was acting like a dam for upstream water. Thoreau likely surmised this, but he did not say so explicitly. Coincidentally, Thoreau's monitoring of Francis's hydraulic experiments on the Concord River coincided with threats of violence against Francis's monitoring on the Lake Village Dam in New Hampshire. Theodore Steinberg, *Nature Incorporated: Industrialization and the Waters of New England* (Cambridge: Cambridge University Press, 1991), 100.

20. Here: Aug. 14, 1859, PT, 24.

21. Each day: Aug. 14, 1859, PT, 244. Tide: Ibid. Action: Ibid. Question: Aug. 14, 1859, PT, 246.

22. Riverbanks: Apr. 1, 1860. Parallel lines: Apr. 1, 1858, PT, 238.

23. Billerica: Jul. 25, 1859, PT, 189. This two-day time for channel transit is a minimum because it does not take into account the delays caused by the Assabet piling up water in the middle of the valley The calculation is: "100 feet in 4½ minutes = 1 mile in 4 hours 100 feet in ¾ minutes = 1 mile in 40 minutes then at 3 miles per hour the water would be 2 days & nights or 48 hours in gaining 16 miles—or to the mouth of the river."

24. Evaporate: Mar. 18, 1859, PT, 58. Season: Aug. 3, 1859, PT, 218. Day: June 23, 1860, PT, 371.

25. Rivers: Richard J. Chorley, A. J. Dunn, and R. P. Beckinsale, *The History of the Study of Landforms, or the Development of Geomorphology* (London: Methuen, 1964), 89. Browsing: Robert Sattelmeyer, *Thoreau's Reading: A Study in Intellectual History (with Bibliographic Catalogue)* (Princeton, NJ: Princeton University Press, 1988), 170, entry 441, checked out from Harvard Library Aug. 15, 1859.

26. Edition: Hunter Rouse and Simon Ince, *History of Hydraulics* (Ames: Iowa Institute of Hydraulic Research, State University of Iowa, 1957), 129. See Kenneth Cameron, "Thoreau's Notes from Dubuat's Principles," *Emerson Society Quarterly* 22 (1961): 68–76.

27. Engineers: Chorley, Dunn, and Beckinsale, *Landforms*, 88. Channels: Ibid, 89. Equilibrium: More technically, "the relation between velocity of flow and resistance. . . . The speed of equilibrium will vary with the nature of the bed" (Pierre Louis Georges Du Buat, *Principes d'hydraulique et de pyrodynamique: vérifiés par un grand nombre d'expériences faites par ordre du gouvernement* [Paris: F. Didot, 1816], 1:110). For example: "If equilibrium is broken, the river strives to re-assert it" (1:117). "The windings of rivers are means used by nature to hasten the establishment of [equilibrium] grade, in spite of the apparent excess of the gradient" (1:150).

28. Swiftness: Du Buat, *Principes*, 82. Du Buat's other materials, or close analogues for them, could all be found in the Concord River: "argile brune" (brown clay), "sable anguleux" (angular sand), "gravier de la Seine" (three textures of gravel from the river Seine), "galets arrondis" (rounded pebbles), and "silex anguleux" (angular stone, i.e., flint the size of an egg). Trapezium: Kenneth W. Cameron, "Thoreau's Notes from Dubuat's Principles," *Emerson Society Quarterly* 22 (1961): 72.

29. Saltation: "A grain of sand, pushed by the current, mounts the gentle slope of the 1st talus, and having arrived at the summit, it rolls by its own weight from the top to the bottom of the opposite talus; there it remains at rest, sheltered from the action of the fluid, and is covered by other grains which come in their turn." Du Buat, *Principes*, 100. On channel curves, he writes: "The bed of the river formed by nature is a curve [courbe] which, on departing from the middle of the current, rises insensibly more & more towards the shores, in proportion

to the steepness of its slopes [pentes], the diminution of the velocities of the strands of water, and to the energy of the molecules of the bed which are to resist it." Du Buat, *Principes,* 118.

30. *Appleton's: Appleton's Dictionary of Machines, Mechanics, Engine-work and Engineering,* 2 vols. (New York: D. Appleton, 1857–1858), cited in Sattelmeyer, *Reading,* 122, entry 57 (Notes in Commonplace Book 2).

31. Thoreau's motive for this analytical work was journaled on Mar. 7, 1859, PT, 227.

32. Linkages: Jan. 24, 1858. Limit: Feb. 28, 1855. New ravine: March 19, PT 276.

33. Eats into: Mar. 7, 1855.

34. Downstream: Aug. 3, 1859, PT, 218. Other side: Jul. 19, 1859, PT, 160. After it: Ibid. Disturbance: July 21, 1859, PT, 168.

35. Channel: Aug. 5, 1858. Material: Jul. 19, 1859, PT, 160. Water: Aug. 1, 1859, PT, 214. Life: Jul. 10, 1859, PT, 140 and Apr. 3, 1859, PT, 112. Brook: Mar. 26, 1860, PT, 114.

36. Vitriol: *Boston Daily Traveller,* Aug. 17, 1859: arguing that a "want of foresight and ignorant and pernicious legislation are the causes of our lands being in their present condition."

37. Drought: Aug. 21, 1859, PT, 247. Life: Aug. 22, 1859, PT, 248.

11. Saving the Meadows

1. Billerica: JCR, 50: "present as Counsel for Petitioners, Judge Mellen, Judge French of New Hampshire, and Mr. Child. For Remonstrants, Judge Abbott, and Messrs. Preston and Somerby. Mr. Butler . . . party interested, and not as counsel." Barrett's Bar: JCR, 51 (below Assabet Bar).

2. Agree: JCR, 51.

3. Wealth: Sept. 16, 1859. Slavery: Sept. 22, 1859. Music: Sept. 28, 1859. Niceties: Oct. 4, 1859. Markets: Oct. 17, 1859. Abolition: Oct. 19, 1859. Fit mood: *The Correspondence of Henry David Thoreau,* ed. Walter Harding and Carl Bode (Westport, CT: Greenwood Press, 1974 [orig. 1958]), 557.

4. The specific case was between James Worster and George W. Young. Theodore Steinberg, *Nature Incorporated: Industrialization and the Waters of New England* (Cambridge: Cambridge University Press, 1991), 99–101.

5. Behind: Ibid., 118.

6. Concord: Robert Gross, "'That Terrible Thoreau': Concord and Its Hermit," in *A Historical Guide to Henry David Thoreau,* ed. W. E. Cain (New York: Ox-

ford University Press, 2000), 197. Irish: Ibid. Career: Patrick Chura, *Thoreau the Land Surveyor* (Gainesville, FL: University Press of Florida, 2010), 155.

7. Parker: Oct. 9, 1859, PT 40.

8. Absurdity: JCR, 66–67.

9. Manning: JCR, 76–78. Barrett: JCR, 93. Buttrick: JCR, 103.

10. Owners: JCR, 127. Grass: JCR, 157.

11. Gristmills: JCR, 195.

12. Passed: JCR, 201.

13. Owners: JCR, 206.

14. Authority: JCR, 255. Future: Hunter Rouse and Simon Ince, *History of Hydraulics* (Ames: Iowa Institute of Hydraulic Research, State University of Iowa, 1957), 165.

15. Inches: More accurate work from 1860 showed this to be a minimum value. It was closer to 8½ inches. Francis: JCR, 255.

16. River: JCR, 268.

17. Feet: JCR, 311. Element: JCR, 317. First mention of hydraulics in the investigation via the Daubison techniques.

18. Case: JCR, 50. Arguments: JCR, 338. Abbott, for the respondents, claimed that the dam was never raised, that it had no impact on the meadow, and that increased in summer wetness was due to the release of water from upstream reservoirs and to the deteriorating quality of the meadows caused by the loss of nutrient. Mellen, for the petitioners, claimed that the dam of 1798 was raised about a foot above its previous level; that the dam of 1828 was raised about 26 inches above the 1798 level, meaning a raising of 37 to 38 inches total above the pre-canal level; and that the deterioration of the meadows was due largely to the increased wetness caused by the dam. Adjourn: JCR, 370.

19. Doubt: JCR, 52.

20. Observations: Dec. 28, 1859, PT, 207.

21. Approval: F. B. Sanborn, letter to Theodore Parker, March 11, 1860, CFPL, cited in Walter Harding, *The Days of Henry Thoreau: A Biography* (New York, Dover, 1982 [orig. 1965]), 429 n. 9.

22. Grass there: Feb. 17, 1860, PT, 6. Ignorance: Mar. 5, 1860, PT, 45.

23. Despotism: JCR, lxxxix. Dead Sea: JCR, xciv.

24. Hearings: J. S. Keyes, *Autobiography* (Concord, MA: n.d.), Concord Free Public Library, 211. "I brought a 'quo warranto' in the Atty Gen name in the S.J.C. against the old Middlesex Canal Co and got their charter forfeited. Then I drew the bill for the relief of the meadow owners by taking down the Billerica dam" (ibid.).

25. Wharf: April 24, 1860, PT, 144. He must have pounded that stake down under the water because his initial reading was "S + 1¼." Its top was also "ten and a half inches below the stone wharf there." By the next afternoon, the river had fallen to "S + 0," being exactly at summer level. To complement his new 1860 round of stage monitoring, Henry also began a short-lived program to track precipitation. Most is numerical data in the journal manuscripts.

26. Data set: May 6, 1860. Comparatively: May 16, 1860.

27. Worst writing: June 6, 1860, PT, 238–239.

28. Lake: TP, Cochituate report

29. Percent of stage: TP. To get a river channel discharge estimate, he used the average of six current velocity measurements and representative cross sections. This he compared with the published releases (discharges) from the compensating reservoirs. The reservoir discharges were very big numbers as cumulative totals. Others calculated the ratio between summer water going to Boston and summer water going down the Sudbury from the upstream compensating reservoir. Only Thoreau converted the total outlet discharge into a measured stream discharge.

30. Low: June 10, 1860, PT, 246.

31. Record: June 21, 1860, PT, 268.

32. Notes on: Harding, *Days,* 440. Friendship: Ibid., 431. Wild Fruits: Henry David Thoreau, *Wild Fruits: Thoreau's Rediscovered Last Manuscript,* ed. Bradley P. Dean. (New York: Norton, 2000).

33. Located etc.: July 12, 1857.

34. Year: June 30, 1860, PT, 286.

35. Systematic: July 7, 1860, PT, 296. Uniformity: Ibid., 297. "June 30th, July 3d, 4th, 6th, and 7th, I carried round a thermometer in the afternoon and ascertained the temperature of the springs, brooks, etc. . . . the average temperature of seventeen is 49½° . . . they do not differ more than 2° from one another. On the whole, then, where I had expected to find great diversity I find remarkable uniformity. . . . This is very near the mean annual temperature of the air here." Climate: July 7, 1860, PT, 298. Phenology: Richard B. Primack, *Walden*

Warming: Climate Change Comes to Thoreau's Woods (Chicago: University of Chicago Press, 2014).

36. Uplands: July 7, 1860, PT, 303. Origin: Ibid. "The average temperature of river [is] more than that of the brooks tried. As the brooks are larger they approach nearer to the river in temperature."

37. Deeper: Aug. 24, 1860, PT, 90.

12. Reversal of Fortune

1. State law: Massachusetts General Court, *Acts and Resolves Passed by the General Court of Massachusetts for the Year 1859, First Session* (Boston: William White, Printer to the State, 1860). Stopped: "Flowages of Land," *New England Farmer* 13 (1861): 162.

2. I knew: J. S. Keyes, *Autobiography* (Concord, MA: n.d.), Concord Free Public Library, 180.

3. "Flowages of Land," 162.

4. Lowermost: Sept. 9, 1860, PT, 118.

5. Directions: Sept. 13, 1860, PT, 83.

6. Change: Oct. 22, 1860, PT, 164.

7. Constitutional: Brian Donahue, "Dammed at Both Ends and Cursed in the Middle: The 'Flowage' of the Concord River Meadows, 1798–1862," *Environmental Review* 13 (1989): 56, citing *Decision of the Supreme Court of Massachusetts upon the Constitutionality of the Act of 1860* . . . (as reported in "Monthly Law Reporter," *Boston*, 1861, 36).

8. Justly: *Statement to the Public in Reference to the Act of the Legislature to Remove the Dam across the Concord River, at Billerica* (Lowell, MA: Stone and Huse, 1860).

9. Living: Richard White, "'Are You an Environmentalist or Do You Work for a Living?': Work and Nature," in *Uncommon Ground: Rethinking the Human Place in Nature*, ed. W. Cronon (New York: Norton, 1995), 171–185.

10. Hypothermic: Walter Harding, *The Days of Henry Thoreau: A Biography* (New York, Dover, 1982 [orig. 1965]), 411. Dec 4, 1860, PT, 41.

11. Governor: Banks was a Boston lawyer who worked as a machinist and as a journalist before returning to school to study law. He was reelected to the governorship in 1859 and 1860, leaving office on January 2, 1861, to serve in the Civil War. Andrews was a Boston lawyer and a former legislator. Elected governor on November 6, 1860, he received the largest majority in state history up to that time. Sickness: Jan. 3, 1861, PT, 307.

12. Beautiful: Jan. 3, 1861, PT, 54–55. Born and dwell: Mar. 28, 1858.

13. Float: Jan. 3, 1861, PT, 55.

14. Step: Jan. 14, 1861, PT, 64. Pigeon: Sept. 12, 1851.

15. Habitat: March 11, 1861, PT, 84.

16. Snow: Apr. 6, 1861, PT, 94. Gossamer: Apr. 6 and 7, 1861, PT, 94–95.

17. Suspend: Massachusetts General Court, *Acts and Resolves Passed by the General Court of Massachusetts for the Year 1861* (Boston: William White, Printer to the State, 1861), 467.

18. Cochituate: John Bogart, *Transactions of the American Society of Civil Engineers, Index Volumes I to XXI Inclusive* (New York: ASCE, 1890), 1–34. Links Shedd with James Francis and with development of the Sudbury River.

19. Report: The report, with a complete Table A, is in the volume *Legislative Documents, House* 1862, 1–150, shelved in Special Collections, in the basement of the Massachusetts State House. The profile uses a zero datum 10.00 feet below "top of bolt." The bolt was placed in 1825 with its top even with the top of the flashboards. The top is 108.6 feet +/− 0.14 feet, North American Vertical Datum of 1988 (el. 109.4 feet +/−, sea level datum of 1929), based on the dam elevation in an October 2000 survey for a Billerica flood insurance study and the 0.72 foot flashboard height in the plan of the dam, ocm31854189.

20. Overcoats: Keyes, *Autobiography*, 188. Duck: Apr. 16, 1861, PT, 97. Hermit thrush: Apr. 20, 1861, PT, 97. Hawk: Apr 25, 1861, PT, 98. Myrtle: May 4 (or 5), 1861, PT, 98.

21. Minnesota: Corinne Hosfeld Smith, *Westward I Go Free: Tracing Thoreau's Last Journey* (Sheffield, VT: Green Frigate Books, 2012). Mississippi: Henry David Thoreau, *The Correspondence of Henry David Thoreau*, ed. Walter Harding and Carl Bode (Westport, CT: Greenwood Press, 1974 [orig. 1958]), 618.

22. Wagon: Thoreau, *Correspondence*, 628.

23. Report: Daniel W. Alvord, Charles S. Storrow, and Hervert J. Shedd, *Report of Experiments and Observations on the Concord and Sudbury Rivers in the Year 1861* (Boston: William White, Printer to the State, 1862).

24. Cause: ECR, 29. Relief: Ibid.

25. Ignored: The river profile drawn for November 18, 1859, on Plate 1 of the full engineering report, and the numerical data (page 106) from which that profile was drawn, in Appendix, General Observations, of Alvord, Storrow, and Shedd, *Engineering Commission Report* (ECR).

26. Data from the ECR show that the dam back-flows the Fordway without covering it. The thalweg of the Fordway is about two feet below the base of the bolt. This natural channel is back-flooded by the dam at the level of 16.5 inches below the top of the bolt.

27. The horizontal distances are given in Alvord, Storrow, and Shedd, *Report,* on page 113, Table of Distances, but the total "fall" was not given.

28. Agency: Massachusetts General Court, *Acts and Resolves . . . 1861,* 273.

29. Skating: Thoreau, *Correspondence,* 634. Edition Walden: Harding, *Days,* 456–457. Edition Week: Ibid, 458.

30. Harding and Bode, *Correspondence,* 641.

31. Edward W. Emerson, *Thoreau: As Remembered by a Young Friend* (Boston: Houghton Mifflin, 1917; reprint, Concord, MA: Thoreau Foundation, 1968), 148n.

32. Repealed: *Massachusetts Acts and Resolves for the year 1861,* 92. "Flowages of Land," 254.

33. Poem: P2, 5, 64, Berg manuscript. Smacking: May 8, 1854.

Conclusion

1. Millers: Apr. 11, 1861, PT, 337.

2. Bring it down: His earlier views foreshadowed Edward Abbey's *The Monkey Wrench Gang* (New York: Avon Books, 1975).

3. Repairs: July 31, 1859, PT 211. Not lie: May 17, 1860, PT, 194.

4. 10 miles: July 22, 1859, PT, 181.

5. Meadows: Constraints are Thoreau's stage monitoring and flood hydrograph analysis from 1860 and discharge calculations from 1859.

6. Clouds: Jan, 1, 1852. Burned: Oct. 10, 1857. Where he is: Nov. 28, 1859. Minimum: For statistics, see David Foster et al., *Wildlands and Woodlands: A Vision for the New England Landscape* (Cambridge, MA: Harvard University Press, 2010). Hoot in: Thoreau was also following the pioneering forest conservation work of George Emerson, *A Report on the Trees and Shrubs Growing Naturally in the Forests of Massachusetts* (Boston: Massachusetts Zoological and Botanical Survey, 1846).

7. Brooks: Oct. 31, 1850.

8. Annually: Mar. 19, 1859, PT, 277. Rods length: Apr. 3, 1859, PT, 336.

9. Tamed: Jan. 28, 1853.

10. Surges: Josiah G. Abbott, *Argument of Hon. Josiah G. Abbott, on Behalf of the Petitioners before the Joint Special Committee of the Legislature of Massachusetts . . . 1862*, 47. Planks: May 4, 1856. Fenny: Sept. 4, 1851.

11. Impetuosity: July 20, 1859, PT, 165.

12. Religion: July 2, 1859, PT, 165.

13. Left: Mar. 23, 1853.

14. Predecessors: James H. Patric and Ernest M. Gould, "Shifting Land Use and the Effects on River Flow in Massachusetts," *Journal of the American Water Works Association*, 1976, 41–45.

15. Might sail: Nov. 7, 1851. My origin: Ibid.

16. John M. Barry, *Rising Tide: The Great Mississippi Flood of 1927 and How It Changed America* (New York: Simon and Schuster, 1977), 21. See also Henry Petroski, *To Engineer Is Human: The Role of Failure in Successful Design* (New York: Vintage, 1992).

17. West Point: John Whiteclay Chambers, *The North Atlantic Engineers: A History of the North Atlantic Division and Its Predecessors in the U.S. Army Corps of Engineers, 1775–1975* (Washington, DC: U.S. Army Corps of Engineers, 1975), 2. Mission: Charles Ellet, *The Mississippi and Ohio Rivers* (Philadelphia: Lippincott, Grambo, 1853). A. A. Humphreys and H. L. Abbot, *Report upon the Physics and Hydraulics of the Mississippi River* (Washington, DC: U.S. Government Printing Office, 1867).

18. Summary: Ellen E. Wohl, *Disconnected Rivers: Linking Rivers to Landscape* (New Haven, CT: Yale University Press, 2004). U.S. Geological Survey, *River Science at the U.S. Geological Survey* (Washington, DC: National Academies Press, 2007). National Research Council, Committee on Characterization of Wetlands, *Wetlands Characteristics and Boundaries* (Washington, DC: National Academy Press, 1995).

19. Landscapes: Edward Hitchcock and Charles H. Hitchcock, *Elementary Geology: A New Edition* (New York: Ivison, Blakeman, Taylor, 1871), i.

20. History: George P. Merrill, "The First One Hundred Years of American Geology, Contributions to the History of American Geology," in *Report of the U.S. National Museum* (Washington, DC: U.S. Government Printing Office, 1904). Daily Life: Examples of Thoreau's science practice are elaborated in Robert M. Thorson, *Walden's Shore: Henry David Thoreau and Nineteenth Century Science* (Cambridge, MA: Harvard University Press, 2014), 201–229 and 262–270.

21. Equilibrium: Grove K. Gilbert, *Report on the Geology of the Henry Mountains*, United States Geological and Geographical Survey, Rocky Mountain Region

(Washington, DC: U.S. Government Printing Office, 1877). See Nicholas J. Clifford, "River Channel Processes and Forms," in *The History of the Study of Landforms,* ed. T. P. Burt, R. J. Chorley, D. Brunsden, N. J. Cox and A. S. Goudie (Trowbridge, UK: Crowell, 1964), 127. New conditions: Nicholas Clifford, "River Channel Processes and Forms," in *The History of the Study of Landforms: Or the Development of Geomorphology,* vol. 4, *Quaternary and Recent Processes and Forms (1890–1965) and the Mid-century Revolutions,* ed. T. P. Burt, R. J. Chorley, D. Brunsden, N. J. Cox and A. S. Goudie (Trowbridge, UK: Crowell), 217–324, 267. River science: Ibid, 217. Theoretical: William M. Davis, "The geographical cycle" *Geographical Journal* 14 (1899): 481–504. Proper theory: Ibid., 217. Grove K. Gilbert, *The Transportation of Debris by Running Water,* Professional Paper 86 (Washington, DC: United States Geological Survey, 1914). Powell: David Applegate, Ruling the Range: Managing the Public's Resources. In Jill S. Schneiderman, ed., *The Earth around Us: Maintaining a Liveable Planet* (New York, W. H., Freeman, 2000), 127.

22. Thoreau recorded many long-term changes in his journal, for example wildlife: Sept. 8, 1854.

23. Coal: Mar. 18, 1857.

Epilogue

1. Polluted: Richard J. Eaton, *A Flora of Concord. Special Publication No. 4* (Cambridge, MA: Harvard Museum of Comparative Zoology, 1974).

2. Glen Canyon: Edward Abbey, *Desert Solitaire* (New York: Ballantine Books, 1968). The politics of this dam controversy is summarized by Robert Wyss, *The Man Who Built the Sierra Club: A Life of David Brower* (New York: Columbia University Press, 2016).

3. Refuge: U.S. Fish and Wildlife Service. Great Meadows National Wildlife Refuge. 2016. www.fws.govrefuge/Great_Meadows/about.html.

References

Abbey, Edward. *Desert Solitaire*. New York: Ballantine Books, 1968.

——. *The Monkey Wrench Gang*. New York: Avon Books, 1975.

Abbott, Josiah G. *Argument of Hon. Josiah G. Abbott, on Behalf of the Petitioners before the Joint Special Committee of the Legislature of Massachusetts, on the Petition of C. P. Talbot and Others, Praying for the Repeal of the Act of 1860 for the Removal of the Dam across Concord River, at Billerica, March 13, 1862*. Boston: Wright and Potter, 1862.

Alvord, Daniel W., Charles S. Storrow, and Herbert J. Shedd. *Report of Experiments and Observations on the Concord and Sudbury Rivers in the Year 1861*. Boston: William White, Printer to the State, 1862.

Applegate, David. "Ruling the Range: Managing the Public's Resources." In *The Earth Around Us: Maintaining a Liveable Planet*, edited by Jill S. Schneiderman, 122–135. New York: W. H. Freeman, 2000.

Appleton's Dictionary of Machines, Mechanics, Engine-work and Engineering. New York: D. Appleton, 1857–1858.

Baron, William R. "Historical Climates of the Northeastern United States: Seventeenth through Nineteenth Centuries." In *Holocene Human Ecology in Northeastern North America*, edited by George P. Nicholas, 29–46. New York: Plenum Press, 1988.

Barry, John M. *Rising Tide: The Great Mississippi Flood of 1927 and How It Changed America*. New York: Simon and Schuster, 1977.

Bartlett, George B., ed. *The Concord Guide Book*. Illustrated by L. B. Pumphrey and Robert Lewis. Boston: Lothrop, 1880.

Blancke, Shirley. "The Archaeology of Walden Woods." In *Thoreau's World and Ours: A Natural Legacy*, edited by Edmund A. Schofield and Robert C. Barron, 242–253. Golden, CO: North American Press, 1993.

Blancke, Shirley, and Barbara Robinson. *From Musketaquid to Concord: The Native and European Experience*. Concord, MA: Concord Antiquarian Museum, 1985.

Board of Cochituate Water Commissioners. *The Overflow of the Meadows of Sudbury River—Report from the Board of Cochituate Water Commissioners.* Boston, 1859.

Bogart, John. *Transactions of the American Society of Civil Engineers, Index Volumes I to XXI Inclusive.* New York: ASCE, 1890.

Boston Daily Traveler, August 17, 1859. In Thoreau's Papers, Concord Free Public Library.

Boston Journal. "The Meadow Lands of the Concord River Valley; Meeting of the Proprietors at Concord." *New England Farmer* 11 (1859): 76–78.

Bridge, John S. *Rivers and Floodplains: Forms, Processes, and Sedimentary Record.* Oxford, UK: Blackwell, 2003.

Buell, Lawrence. *The Environmental Imagination.* Cambridge, MA: Harvard University Press, 1995.

Cain, William E. Henry David Thoreau, "1817–1862: A Brief Biography." In *A Historical Guide to Henry David Thoreau,* edited by W. E. Cain, 1–11. New York: Oxford University Press, 2000.

Cameron, Kenneth W. "Thoreau's Notes from Dubuat's Principles." *Emerson Society Quarterly* 22 (1961): 68–76.

———. "Thoreau in the Court of Common Pleas." *Emerson Society Quarterly* 14 (1959): 86.

Cameron, Sharon. *Writing Nature: Henry Thoreau's Journal.* New York: Oxford University Press, 1985.

Caradonna, Jeremy L. 2014. *Sustainability: A History.* New York: Oxford University Press, 2014.

Chambers, John Whiteclay. *The North Atlantic Engineers: A History of the North Atlantic Division and Its Predecessors in the U.S. Army Corps of Engineers, 1775–1975.* Washington, DC: U.S. Army Corps of Engineers, 1975.

Channing, William Ellery. *Thoreau the Poet Naturalist with Memorial Verses.* New edition. Edited by F. B. Sanborn. Boston, Charles Goodspeed, 1902.

Chorley, Richard. J., A. J. Dunn, and R. P. Beckinsale. *The History of the Study of Landforms: or the Development of Geomorphology.* Vol. 1 of *Geomorphology before Davis.* London: Methuen, 1964.

Chura, Patrick. *Thoreau the Land Surveyor.* Gainesville: University Press of Florida, 2010.

Clifford, Nicholas J. "River Channel Processes and Forms." In *The History of the Study of Landforms: Or the Development of Geomorphology*. Vol. 4 of *Quaternary and Recent Processes and Forms (1890–1965) and the Mid-century Revolutions*, edited by T. P. Burt, R. J. Chorley, D. Brunsden, N. J. Cox and A. S. Goudie, 217–324. Trowbridge, UK: Crowell, 1964.

"Complaint and Petition of the Inhabitants of Wayland, Sudbury, Concord, Bedford and Carlisle." Boston: Alfred Mudge and Son, 1859. Concord Free Public Library Pamphlet C.PAM.60 Item B1.

Cronon, William. *Changes in the Land: Indians, Colonists, and the Ecology of New England*. New York: Hill and Wang, 1983.

———. "The Trouble with Wilderness; or, Getting Back to the Wrong Nature." In *Uncommon Ground: Rethinking the Human Place in Nature*, edited by W. Cronon, 69–90. New York: W. W. Norton, 1996.

Crutzen, P. J., and E. F. Stoermer. "The Anthropocene." *Global Change Newsletter* 41 (2000): 17–18.

Cumbler, John. *Reasonable Use: the People, Environment, and the State, 1770–1930*. Oxford, UK: Oxford University Press, 2001.

"The Dam at North Billerica." *New England Farmer* 14 (1862): 254–255.

D'Arcy Wood, Gillen. *Tambora: The Year That Changed the World*. Princeton, NJ: Princeton University Press, 2013.

Darwin, Charles R. *The Voyage of the Beagle*. Edited by Charles W. Eliot. New York: P. F. Collier and Son, 1963. Originally published 1845.

Davis, William M. "The Geographical Cycle." *Geographical Journal* 14 (1899): 481–504.

Dean, Bradley P. Introduction. In Henry David Thoreau, *Wild Fruits: Thoreau's Rediscovered Last Manuscript*. Edited by Bradley P. Dean. New York: Norton, 2000.

Deetz, James. *In Small Things Forgotten: An Archaeology of Early American Life*. New York: Doubleday, 1977.

Demos, John. *Circles and Lines: The Shape of Life in Early America*. Cambridge, MA: Harvard University Press, 2004.

Donahue, Brian. "Dammed at Both Ends and Cursed in the Middle: The 'Flowage' of the Concord River Meadows, 1798–1862." *Environmental Review* 13 (1989): 46–67.

———. *The Great Meadow: Farmers and the Land in Colonial Concord*. New Haven, CT: Yale University Press, 2004.

Du Buat, Pierre Louis Georges. *Principes d'hydraulique et de pyrodynamique: véri-fiés par un grand nombre d'expériences faites par ordre du gouvernement.* 3 vols. Paris: F. Didot, 1816.

Eaton, Gordon. "Down to Earth: A Historical Look at Government-Sponsored Geology." In *The Earth around Us: Maintaining a Liveable Planet,* edited by Jill S. Schneiderman, 83–98. New York: W. H. Freeman, 2000.

Eaton, Richard J. *A Flora of Concord.* Special Publication No. 4. Cambridge, MA: Harvard Museum of Comparative Zoology, 1974.

Edwards, L.E. "What is the Anthropocene?" *EOS* 97 (2015): 6–7.

Ellet, Charles. *The Mississippi and Ohio Rivers.* Philadelphia: Lippincott, Grambo, 1853.

Emerson, Edward W. *Thoreau: As Remembered by a Young Friend.* Boston: Houghton Mifflin, 1917. Reprint: Concord, MA: Thoreau Foundation, 1968.

Emerson, George B. *A Report on the Trees and Shrubs Growing Naturally in the Forests of Massachusetts.* Boston: Dutton and Wentworth, 1846.

Emerson, Ralph W. *The Journals and Miscellaneous Notebooks of Ralph Waldo Emerson,* vol. 14, *1854–1861.* Edited by S. S. Smith and H. Hayford. Cambridge, MA: Belknap Press of Harvard University Press, 1978.

———. *The Letters of Ralph Waldo Emerson,* vol. 8, *1845–1859.* Edited by Eleanor M. Tilton. New York: Columbia University Press, 1991.

———. *The Letters of Ralph Waldo Emerson,* vol. 9, *1860–1869.* Edited by Eleanor M. Tilton. New York: Columbia University Press, 1991.

———. *Nature.* Boston: James Munroe, 1836.

———. "Thoreau." In Henry David Thoreau, *Excursions,* edited by Leo Marx, 7–33. Boston: Ticknor and Fields, 1863.

Finney, Stanley C. and L. E. Edwards. "The 'Anthropocene' Epoch: Scientific Decision or Political Statement?" *GSA Today* 26 (2016): 4–10.

Foley, Robert. *Paint in 18th-Century Newport.* Newport, RI: Newport Restoration Foundation, 2016.

Foster, David, et al. *Thoreau's Country: A Journey through a Transformed Landscape.* Cambridge, MA: Harvard University Press, 1999.

———. *Wildlands and Woodlands: A Vision for the New England Landscape.* Cambridge, MA: Harvard University Press, 2010.

French, Henry F. *Argument of Hon. Henry F. French of Boston, March 12, 1862, before the Joint Committee of the Legislature of Mass. on the Expedition for the Repeal of an Act in Relation to the Flowage of the Meadows on Concord and Sudbury Rivers. Approved April 4, 1860. Much Condensed.* Boston. Concord Free Public Library pamphlet C. PAM.60 Item B3, 1862.

Gilbert, Grove K. *Report on the Geology of the Henry Mountains.* United States Geological and Geographical Survey, Rocky Mountain Region. Washington, DC: U.S. Government Printing Office, 1877.

———. *The Transportation of Debris by Running Water.* Professional Paper 86. Washington, DC: United States Geological Survey, 1914.

Goldthwait, J. Walter. "The Sand Plains of Glacial Lake Sudbury." *Bulletin of the Museum of Comparative Zoology at Harvard College* 42 (1905).

Gould, Stephen J. *Bully for Brontosaurus: Reflections on Natural History.* New York: Oxford University Press, 2000.

Grahame, Kenneth. *The Wind in the Willows.* New York: Charles Scribner, 1908.

Gray, Horace, Jr. *Reports: Supreme Judicial Court of Massachusetts, 1871.*

Grayson, Stan. "The Musketaquid Mystery: In Search of Thoreau's Boat." *Wooden Boat Magazine* 186 (2005): 39–45.

Gross, Robert G. "'Faith in the Boardinghouse': New Views of Thoreau Family Religion." *Thoreau Society Newsletter* 250 (Winter 2005): 1–5.

———. *The Minutemen and Their World.* New York: Hill and Wang, 1976.

———. "The Most Estimable Place in All the World: A Debate on Progress in Nineteenth-Century Concord." *Studies in the American Renaissance,* 1978, 1–15.

———. "'That Terrible Thoreau': Concord and Its Hermit." In *A Historical Guide to Henry David Thoreau,* edited by W. E. Cain, 181–242. New York: Oxford University Press, 2000.

Guldi, Jo, and David Armitage. *The History Manifesto.* New York: Cambridge University Press, 2014.

Harding, Walter. *The Days of Henry Thoreau: A Biography.* New York, Dover, 1982. Originally published 1965.

———. "Foreword." In Henry David Thoreau, *The Journal of Henry David Thoreau,* edited by Bradford Torrey and Francis H. Allen, v–vii. New York: Dover Publications, 1984. Originally published 1906.

Hawthorne, Nathaniel. *The American Notebooks*. New Haven, CT: Yale University Press, 1932.

———. *Mosses from an Old Manse*. London: Wiley and Putnam, 1846.

———. *Passages from the American Notebooks, Volume 2*. Chapel Hill, NC: Project Gutenberg EBook, 2005.

Hazen, Henry A. *History of Billerica, Massachusetts, with a Genealogical Register*. Boston: A. Williams, 1883.

Henderson, David. "American Wilderness Philosophy." *The Internet Encyclopedia of Philosophy*. Available at http://www.iep.utm.edu/am-wild/(accessed November 10, 2016).

"Henry David Thoreau (1817–1862) Herbarium, Container List." Harvard University Herbaria. http://botlib.huh.harvard.edu/libraries/Thoreau_container.htm

Hitchcock, Edward. *Final report on the Geology of Massachusetts in Four Parts*. Northampton: J. Butler, 1841.

Hitchcock, Edward, and Charles H. Hitchcock. *Elementary Geology: A New Edition*. New York: Ivison, Blakeman, Taylor, 1871.

Horton, Robert E. *Weir Experiments, Coefficients, and Formulas*. Revision of paper no. 150. Washington, DC: U.S. Government Printing Office, 1907.

Horwitz, Morton J. *The Transformation of American Law, 1780–1860*. Cambridge, MA: Harvard University Press, 1977.

Howarth, William L. "Concord River Survey (F13)." In *The Literary Manuscripts of Henry David Thoreau*, edited by W. L. Howarth, 303. Columbus: Ohio State University Press, 1974.

Humphreys, A. A., and H. L. Abbot. *Report upon the Physics and Hydraulics of the Mississippi River*. Washington, DC: U.S. Government Printing Office, 1867.

Judd, Richard. *Second Nature: An Environmental History of New England*. Amherst: University of Massachusetts Press, 2014.

Keyes, J. S. *Autobiography*. Concord, MA, n.d. Concord Free Public Library.

Koteff, Carl. *Glacial Lakes Near Concord, Massachusetts*. Professional Paper 475-C. Washington, DC: U.S. Geological Survey, 1963.

Koteff, Carl, G. R. Robinson, R. Goldsmith, and W. B. Thompson. "Delayed Postglacial Uplift and Synglacial Sea Levels in Coastal Central New England." *Quaternary Research* 40 (1993): 46–54.

Leff, David. *Deep Travel: In Thoreau's Wake on the Concord and Merrimack.* Ames: University of Iowa Press, 2009.

Linklater, Andro. *Measuring America: How the United States Was Shaped by the Greatest Land Sale in History.* New York: Penguin, 2003.

Luria, Sara. "Thoerau's Geopoetics." In *GeoHumanities: Art, History, Text at the Edge of Place,* edited by M. Dear, J. Ketchum, S. Luria, and D. Richardson, 126–138. London: Routledge, 2011.

Mann, Charles C. *1491: New Revelations of the Americas before Columbus.* New York: Alfred A. Knopf, 2005.

Marsh, George P. *Man and Nature or, Physical Geography as Modified by Human Action.* New York: Charles Scribner, 1864.

Marx, Leo. *The Machine in the Garden: Technology and the Pastoral Ideal in America.* New York: Oxford University Press, 1964.

———. "Walden as Transcendental Pastoral Design." In *Walden and Resistance to Civil Government: Authoritative Texts, Thoreau's Journal, Reviews an Essays in Criticism,* 2nd ed., edited by William Rossi, 377–390. New York: W. W. Norton, 1992.

Massachusetts General Court. *Acts and Resolves Passed by the General Court of Massachusetts for the Year 1859, First Session.* Boston: William White, Printer to the State, 1860.

———. *Acts and Resolves Passed by the General Court of Massachusetts for the Year 1860.* Boston: William White, Printer to the State, 1861.

———. *Acts and Resolves Passed by the General Court of Massachusetts for the Year 1861.* Boston: William White, Printer to the State, 1861.

Massachusetts Joint Special Committee. *Report of the Joint Special Committee Upon the Subject of The Flowage of Meadows on Concord and Sudbury Rivers, January 28, 1860.* Boston: William White, Printer to the State, 1860.

Maynard, Barksdale W. *Walden Pond, a History.* New York: Oxford University Press, 2004.

McKibben, Bill. *The End of Nature.* New York: Random House, 1989.

McLean, Albert F., Jr. Addenda to the Thoreau Correspondence. *Bulletin of the New York Public Library* 71 (1967): 265–267.

McPhee, John. *The Founding Fish.* New York: Farrar, Straus and Giroux, 2003.

———. "Introduction." In Henry David Thoreau, *A Week on the Concord and Merrimack Rivers,* edited by C. F. Hoyde, W. L. Howarth, and E. H. Witherell, ix–xlvi. Princeton, NJ: Princeton University Press, 2004.

Merrill, George P. "The First One Hundred Years of American Geology: Contributions to the History of American Geology." *Report of the U.S. National Museum under the Directorate of the Smithsonian Institution for the Year Ending June 30, 1904.* Washington, DC: U.S. Government Printing Office, 1904.

Miller, Daegan R. "Witness Tree: Landscape and Dissent in the Nineteenth-Century United States." Ph.D. dissertation, Cornell University, 2013.

Mock, Cary J., Jan Mojzisek, Michele McWaters, Michael Chenoweth, and David W. Stahle. "The Winter of 1827–1828 over Eastern North America: A Season of Extraordinary Climatic Anomalies, Societal Impacts, and False Spring." *Climate Change* 83 (2007): 87–115.

Montgomery, David. *King of Fish: The Thousand-Year Run of Salmon.* Cambridge, MA: Perseus Books, 2003.

Muir, Diana. *Reflections in Bullough's Pond: Economy and Ecosystem in New England.* Hanover, NH: University Press of New England, 2000.

Naiman, Robert J., Carol A. Johnston, and James C. Kelley. "Alteration of North American Streams by Beaver." *Bioscience* 38 (1988): 753–762.

National Research Council, Committee on Characterization of Wetlands. *Wetlands: Characteristics and Boundaries.* Washington, DC: National Academy Press, 1995.

Nye, David E. *America as Second Creation: Technology and Narratives of New Beginnings.* Cambridge, MA: MIT Press, 2003.

Oswald, W. Wyatt, et al. "Post-Glacial Changes in Spatial Patterns of Vegetation across Southern New England." *Journal of Biogeography* 34 (2007): 900–913.

Patric, James H., and Ernest M. Gould. "Shifting Land Use and the Effects on River Flow in Massachusetts." *Journal of the American Water Works Association,* 1976, 41–45.

Peltier, W. R. "Global Glacial Isostasy and the Surface of the Ice-Age Earth: The ICE-5G (VM2) Model and GRACE." *Annual Reviews of Earth and Planetary Science* 32 (2004): 111–149.

Petroski, Henry. *To Engineer Is Human: The Role of Failure in Successful Design.* New York: Vintage, 1992. Originally published 1985.

Pratt, Mark. "Small Part of Massachusetts Said to Be in 'Extreme Drought.'" *Chicago Tribune,* August 11, 2016.

Primack, Richard B. *Walden Warming: Climate Change Comes to Thoreau's Woods.* Chicago: University of Chicago Press, 2014.

Prude, Jonathan. *The Coming of Industrial Order: Town and Factory Life in Rural Massachusetts, 1810–1860,* 2nd ed. New York: Cambridge University Press, 1999.

Purdy, Jedediah. *After Nature: A Politics for the Anthropocene.* Cambridge, MA: Harvard University Press, 2014.

Rawson, Michael. *Eden on the Charles The Making of Boston.* Cambridge, MA: Harvard University Press, 2010.

Richardson, Laurence E. *Concord River.* Barre, MA: Barre Publishing, 1964.

Richardson, Robert D., Jr. *Henry Thoreau: A Life of the Mind.* Berkeley: University of California Press, 1986.

———. "Thoreau and Science." In *American Literature and Science,* edited by Robert J. Scholnick, 110–127. Lexington: University Press of Kentucky, 1992.

Rouse, Hunter, and Simon Ince. *History of Hydraulics.* Ames: Iowa Institute of Hydraulic Research, State University of Iowa, 1957.

Ruddiman, W. F., E. C. Ellis, J. O. Kaplan, and D. Fuller. "Defining the Epoch We Live In." *Science* 348 (2015): 38–39.

Ruland, Richard, and Malcom Bradbury. *From Puritanism to Postmodernism: A History of American Literature.* New York: Viking, 1991.

Sanborn, Frank. B. *Hawthorne and His Friends: Reminiscence and Tribute.* Cedar Rapids, IA: Torch Press, 1908.

———. *Henry D. Thoreau.* Boston: Houghton, Mifflin, 1882. Reprint: New York: Chelsea House, 1980.

———. *The Life of Henry David Thoreau: Including Many Essays Hitherto Unpublished, and Some Account of His Family and Friends.* Boston: Houghton Mifflin, 1917.

Sattelmeyer, Robert. *Thoreau's Reading: A Study in Intellectual History (with Bibliographic Catalogue).* Princeton, NJ: Princeton University Press, 1988.

Shattuck, Lemuel. *History of the Town of Concord; Middlesex County, Massachusetts, from Its Earliest Settlement to 1832.* Boston: Russell, Odiorne, 1835.

Schultz, Kathryn. "Pond Scum: Henry David Thoreau's Moral Mypoia." *New Yorker,* October 19, 2015.

Sims, Michael. *The Adventures of Henry Thoreau: A Young Man's Unlikely Path to Walden Pond.* New York: Bloomsbury, 2015.

Smith, Corinne Hosfeld. *Westward I Go Free: Tracing Thoreau's Last Journey.* Sheffield, VT: Green Frigate Books, 2012.

"Statement to the Public in Reference to the Act of the Legislature to Remove the Dam across the Concord River, at Billerica." Lowell: Stone and Huse, 1860. Concord Free Public Library Pamphlet C.PAM.60 Item B2.

Stegner, Wallace, *Crossing to Safety,* New York: Modern Library 2002.

Steinberg, Theodore. *Nature Incorporated: Industrialization and the Waters of New England.* Cambridge: Cambridge University Press, 1991.

Stone, Janet R. and Byron Stone. "Surficial Geologic Map of the Clinton-Concord-Grafton-Medfield 12-Quadrangle Area in East Central Massachusetts, Scale 1:24,000." U.S. Geological Survey Open File Report 2006-1260A. DVD. 2006.

Tauber, Alfred I., *Henry David Thoreau and the Moral Agency of Knowing.* Berkeley: University of California Press, 2001.

Thoreau, Henry David. *Cape Cod.* Boston: Ticknor and Fields, 1865.

——. *The Correspondence of Henry David Thoreau.* Edited by Walter Harding and Carl Bode. Westport, CT: Greenwood Press, 1974. Originally published 1958.

——. *Excursions.* Boston: Ticknor and Fields, 1863.

——. *Journal 1: 1837–1844.* Edited by Elizabeth Witherell, William. L. Howarth, Robert Sattelmeyer, and Thomas Blanding. Princeton, NJ: Princeton University Press, 1981.

——. *Journal 2: 1842–1848.* Edited by Robert Sattelmeyer. Princeton, NJ: Princeton University Press, 1984.

——. *Journal 3: 1848–1851.* Edited by Robert Sattelmeyer. Princeton NJ: Princeton University Press, 1990.

——. *The Journal of Henry D. Thoreau: In Fourteen Volumes Bound as Two.* Edited by Bradford Torrey and Francis H. Allen. New York: Dover, 1962. Originally published 1906.

——. Journal Online Transcripts. Manuscripts: Vols. 28, 29, 30, 31, and 32. Edited by Princeton editorial team. Available at http://thoreau.library.ucsb.edu/writings_journals.html.

———. *The Maine Woods.* Boston: Ticknor and Fields, 1864.

———. "Natural History of Massachusetts". *Dial* 3 (1842): 19–40.

———. "Paradise (to Be) Regained." *United States Magazine and Democratic Review* 13 (1843): 451–462.

———. "The Succession of Forest Trees." In *The Succession of Forest Trees and Wild Apples by Henry D. Thoreau with a Biographical Sketch by Ralph Waldo Emerson,* 33–52. Boston: Houghton Mifflin, 1887. Originally published 1860.

———. "Walking." *Atlantic Monthly* 9 (1862): 657–674.

———. *A Week on the Concord and Merrimack Rivers.* Edited by Thomas Blanding. Orleans, MA: Parnassus Imprints, 1987; reprinted, Boston: James Monroe & Co., 1849.

———. "A Winter Walk." *Dial* 4 (1843): 211–226.

Thorson, Robert M. *Walden's Shore: Henry David Thoreau and Nineteenth Century Science.* Cambridge, MA: Harvard University Press, 2014.

Torrey, Bradford and Francis Allen. "Introduction." In *The Journal of Henry David Thoreau in Fourteen Volumes Bound as Two.* New York: Dover, 1962, originally Boston, Houghton Mifflin, 1906, 219.

Traubel, Horace. *With Walt Whitman in Camden.* 3 vols. 1905–1906. Reprint: New York: Rowman and Littlefield, 1961.

U.S. Fish and Wildlife Service. "Great Meadows National Wildlife Refuge." 2016. Available at www.fws.govrefuge/Great_Meadows/about.html.

U.S. Geological Survey. *River Science at the U.S. Geological Survey.* Washington, DC: National Academies Press, 2007.

Waldrop, M. Mitchell. *Complexity: The Emerging Science at the Edge of Order and Chaos.* New York: Touchstone, 1992.

Walling, H. F. *Map of the Town of Concord, Middlesex County Mass.* Boston: No. 31 Washington St., 1852.

Walls, Laura Dassow. *Material Faith: Henry David Thoreau on Science.* Boston: Houghton Mifflin, 1999.

———. *Seeing New Worlds: Henry David Thoreau and Nineteenth-Century Natural Science.* Madison: University of Wisconsin Press, 1995.

White, Richard. "'Are You an Environmentalist or Do You Work for a Living?': Work and Nature." In *Uncommon Ground: Rethinking the Human Place in Nature,* edited by W. Cronon, 171–185. New York: Norton, 1995.

———. *The Organic Machine: The Remaking of the Columbia River.* New York: Hill and Wang, 1995.

Wohl, Ellen E. *Disconnected Rivers: Linking Rivers to Landscape.* New Haven, CT: Yale University Press, 2004.

Wood, William. *New-England's Prospect.* Boston: Prince Society, 1639.

Worster, Donald. *Nature's Economy: A History of Ecological Ideas,* 2nd ed. New York: Cambridge University Press, 1994.

Wyss, Robert. "The Man Who Built the Sierra Club: A Life of David Brower." New York: Columbia University Press, 2016.

Zen E-an et al. *Bedrock Geologic Map of Massachusetts. (Scale 1:250,000).* Washington, DC: U.S. Geological Survey, 1983. http://mrdata.usgs.gov/sgmc/ma.html.

Zwinger, Ann, and Edwin Way Teale. *A Conscious Stillness: Two Naturalists on Thoreau's Rivers.* Amherst: University of Massachusetts Press, 1984.

Acknowledgments

This book would not have been possible without the advice, support, and help of several key people. Kristine, my wife, helped at every stage of this project—sharing her insights, managerial talents, and editorial skills. Leslie Perrin Wilson of the Concord Free Public Library specifically asked me to examine Thoreau's scroll map, and followed up with expert archival support. John Kulka, my editor at Harvard University Press, proposed a follow-up book to our previous collaboration, guided its progress, and suggested the title. Brian Donahue from the Department of History at Brandeis University encouraged the launch of this project and later provided a very helpful review. Lisa Adams of the Garamond Agency was consistently supportive. David Wood of the Concord Museum shared many insights and loaned the *Musketaquid* replica for our frontispiece photograph, which was expertly taken by Juliet Wheeler.

Thanks to the University of Connecticut for supporting my left turn from lab-based physical science to history and biography, and for financial support. My humanities colleagues at UConn, Christopher Clark, Wayne Franklin, and Robert Gross, provided steerage through the river bends of this book. Those in the Center for Integrative Geosciences, Lisa Park Boush, Katherine Johnson, and William Ouimet, shared ideas and technical support. Valori Banfi, Scott Kennedy, and Donovan Reinwald at the Homer Babbidge Library provided research help when I got stuck.

Additional archival assistance was provided by Jennifer Fauxsmith at the Massachusetts State Library, Josh McKeon at the Berg (New York City Public) Library, Jeffrey Cramer at the Thoreau Institute, Conni Manoli of the Concord Free Public Library, and Patricia Gilrein at the Concord Museum. Kenneth Turkington and Ellen Kisslinger provided logistical help on the Concord and Assabet Rivers, respectively. Botanists Robert Capers of the University of Connecticut herbarium and Ray Angelo of the Harvard Department of Organismic and Evolutionary Biology helped answer specific questions. Archaeologists Shirley Blancke and Brian Jones gave advice. Historians Jeffrey Eagan and Theodore Steinberg offered helpful tips. Christine Laudon provided logistical support.

Sherman Clebnik, Robert Gross, David Leff, Ellen Wohl, and Sylvia Jane Wojcik provided helpful comments on an earlier draft of this text. Joy Deng and Julia Kirby of Harvard University Press, Mikala Guyton of Westchester Publishing Services, and Sue Warga gave expert editorial help.

Index